TEXAS AMPHIBIANS

TEXAS NATURAL HISTORY GUIDES™

TEXAS AMPHIBIANS

A FIELD GUIDE

BOB L. TIPTON, TERRY L. HIBBITTS,
TOBY J. HIBBITTS, TROY D. HIBBITTS,
AND TRAVIS J. LADUC

UNIVERSITY OF TEXAS PRESS
Austin

The publication of this book was supported in part by the Corrie Herring Hooks Series Endowment in Natural History.

Unless otherwise indicated, all photos were taken by the authors.

LIBRARY OF CONGRESS CATALOGING-IN-PUBLICATION DATA

Texas amphibians : a field guide / Bob Tipton . . . [et al.]. — 1st ed.
    p.   cm. — (Texas natural history guides)
  Includes bibliographical references and index.
  ISBN 978-0-292-73735-8 (pbk. : alk. paper) — ISBN 978-0-292-73736-5 (e-book)
  1. Amphibians—Texas—Identification.   I. Tipton, Bob L., 1948–
  QL653.T4T49   2012
  597.809764—dc23

                           2011039280

*We dedicate this book to
our good friend and coauthor,
the late Bob L. Tipton,
who was instrumental
in getting us started
on this publication,
and to James R. Dixon,
who has been a mentor, friend,
and fellow field herper to each of us.*

CONTENTS

Contents

FOREWORD

The quest of most naturalists is for a hands-on book about a subject close to their hearts, usually a favorite species or group of species that they wish to learn more about. In Texas, this subject isn't easy to resolve. The state contains at least five biotic provinces and an almost equal number of biomes. Each formation contains a wealth of species, living, dying, day by day, year by year, leaving their legacies for some young naturalist to discover.

In such a diverse biological realm, those of us who study amphibians have a difficult task because we must travel from the swamps of Southeast Texas to the Chihuahuan Desert of West Texas, or from the grassy plains of the Panhandle to the tropics of South Texas to find a salamander or frog we wish to study. Each area of Texas has a particular set of species that have evolved over thousands of years. For example, the Edwards Plateau, where no less than 14–17 species of spring and cave salamanders have evolved, or southern Texas where remnants of Mexican tropical salamanders, frogs, and toads still survive under the strong pressures of human population growth and crop applications of chemical herbicides and pesticides. Timber harvests in East Texas and grain harvests in the Panhandle have depleted the natural occurrence of humidity, water flow, and groundwater to the point of no return. With the continued depletion of land and water resources, along with natural drought conditions, we must ask ourselves, when will it end?

The five authors of this book, *Texas Amphibians*, have asked themselves the above questions and now invite readers to discover what they have learned about the natural history, conservation, and laws concerning the 72 species of Texas frogs, toads, and salamanders.

The authors also ask that concerned citizens be aware of their surroundings and work to save some species from extinction. The way we live our lives can either reflect the continued music of nightly choruses of spring peepers or the harsh reality of Rachel Carson's *Silent Spring*.

JAMES R. DIXON

ACKNOWLEDGMENTS

We would like to express special thanks to our wives—Diana Hibbitts, Marla Hibbitts, Rachel Hibbitts, and Veronica LaDuc—to our daughters—Cheyenne Hibbitts, Holly Hibbitts, and Emma LaDuc—and to our extended families for their support of our interest in nature, especially in reptiles and amphibians, which led to this book. We would especially like to acknowledge Bob's wife, Angie Tipton, and their daughters, Brandy Morgan, Carrie Benefield, and Anna Holt, for their help during his illness, as well as Bob's neighbor Tom Rose for all the reading and editing he did with Bob during this time.

We appreciate James R. Dixon, professor emeritus in the Department of Wildlife and Fisheries Science at Texas A&M University, who, besides writing the foreword to the book, was our support and inspiration and allowed us to use an adapted version of his amphibian keys. We also want to thank John Malone for his input on Texas amphibians, for the use of an adapted version of his larval amphibian key, and for his discussions about *Syrrhophus*. Thanks to Corey Roelke for his input on the South

Texas Siren, and to Brian Fontenot for discussions about toads in East Texas. Andy Gluesenkamp assisted with discussions about endemic Texas *Eurycea* and with information about photographing *Eurycea*. We would also like to thank the owners of Cascade Caverns and Josh Leerhoff (cave manager) for allowing us to photograph salamanders in their cave. Thanks to Jared Holmes for helping us find *Eurycea* in Blanco County. Our reviewers, Robert Hansen and Charles Painter, were invaluable for their suggestions and support of this book. Casey Kittrell of the University of Texas Press was very patient with us and was a big help in getting this book ready for publication.

Additional photographs were graciously supplied by Tim Burkhardt, Danté Fenolio, Joe Furman, Scott Wahlberg, and Kenny Wray. Dee Ann Chamberlain let us photograph some Austin-area salamanders and provided information on other sites where they might be found. Joe Fries and Patricia Grant of the U.S. Fish and Wildlife Service in San Marcos also allowed us to photograph some of their captive Spring Salamanders.

TEXAS AMPHIBIANS

INTRODUCTION

Members of the class Amphibia are collectively known as "amphibians," a term derived from the Greek roots *amphi* and *bios*, literally "life on both sides," that is, life on land and in water. Although there are exceptions, amphibians stereotypically lay unprotected eggs in water, which then develop into aquatic larvae that metamorphose into terrestrial adults. Amphibians may frequently be confused with reptiles; however, unlike reptiles, amphibians have clawless toes and thin, moist, scaleless skins. Although amphibians and reptiles are often studied together in a branch of zoology called herpetology, these 2 groups are not particularly closely related. Amphibians are found worldwide (except in Antarctica), and are classified in 3 orders: Caudata (salamanders), Anura (frogs), and Gymnophiona (caecilians). Approximately 6,800 species have been described worldwide.

The salamanders (order Caudata) are composed of just more than 600 species. Salamanders are distributed primarily across the Northern Hemisphere, occurring mostly in North America,

Europe, and Asia, with a few species reaching the mountains of extreme northern Africa, as well as an important secondary radiation of species into South America. Salamander adults can be distinguished from frogs and toads because they have a slender body with a long tail and (usually) 4 legs of equal length. Most salamanders have true teeth in both jaws. Although salamanders superficially resemble lizards, they lack the claws and scales possessed by lizards and most other reptiles.

Frogs (order Anura) are considered to be the most successful amphibian group, with about 6,000 species worldwide, occurring on all continents except Antarctica. The name "Anura" means "without a tail," a feature common to all species. They are most numerous and diverse in tropical regions. Frogs are some of the most distinctive of vertebrates. No other living group of adult vertebrates has the combination of 4 well-developed limbs, hind limbs enlarged and modified for jumping, the lack of an external tail, and well-developed eyes with lids. Typically, frogs lack teeth in the lower jaws. Frogs also possess mucous and toxin glands in their skin, which keep them moist and protect them from predators, respectively.

The final amphibian order, the caecilians (order Gymnophiona), consists of about 200 species that occur in tropical or semitropical countries and are not found in the United States. These little-known amphibians have a series of annuli, or rings, around their bodies, and they often closely resemble earthworms. Caecilians practice internal fertilization; some species lay eggs, while other are live-bearers. Some species are aquatic throughout their lives, while others leave the water after metamorphosing from their aquatic larval stage and live in burrows in damp soil and leaf litter. Caecilians are so difficult to find and observe that there are many gaps in our knowledge about them, and it is likely that many more species will be discovered.

Ancestral amphibians found in the oldest fossils were apparently totally aquatic, but during the Devonian Period, about 365 million years ago, they made the transition to spending at least part of their lives on land. These ancient amphibians were still closely tied to water because it was required for their reproduction and the survival of their unshelled, permeable eggs. Many modern amphibians retain this primitive, dual life-history strategy of laying eggs that transform into aquatic, externally gilled larvae that later metamorphose into moisture-dependent terrestrial adults.

All amphibians are ectotherms, and unlike birds and mammals, which generate body heat by metabolizing food, they must rely on environmental sources for body heat. While this may seem to be a disadvantage, ectothermy has its benefits. Since amphibians do not use food to generate body heat, they require less food per individual than endotherms. This allows them to survive in areas where there is less food than would be necessary for similarly sized endotherms. Second, it sometimes allows for greater population densities than would be possible for endotherms subsisting on the same food resource. Third, ectothermy allows them to undergo long periods of inactivity without feeding. In fact, some amphibians (for example, spadefoots) may undergo periods of dormancy in excess of one year while they wait for seasonal rainfall and the opportunity to feed and reproduce.

At some stage in their lives, all amphibians exhibit the ancestral characteristic of external gills. In the more ancestral species, gills are present in aquatic larval stages but lost in terrestrial adults. By contrast, many derived species undergo direct development, in which the larval stage is abbreviated and occurs entirely within the egg, with gills never present in free-living individuals. In other derived species, larval characteristics, including external gills, are retained in aquatic adults, which are referred to as "neotenic," "paedomorphic," or "perennibranchiate" because they retain these and other larval morphological features throughout their lives. No other limbed vertebrates have external gills at any stage of their development. Amphibian eggs lack shells and usually can be identified to species only by microscopic examination.

Amphibian larvae may be herbivores, omnivores, or carnivores, depending primarily on specific species adaptations. All amphibians are predatory carnivores as adults, preying mainly on invertebrates, although certain large species, such as American Bullfrogs and amphiumas, may feed on relatively large vertebrate prey. The environmental benefits from the vast numbers of invertebrates consumed daily by amphibians are incalculable.

Globally, many (perhaps most) populations of amphibians are in decline. In addition to the problems of habitat loss from human activity, which affects all animals, amphibian populations are uniquely susceptible to declines because of their unique natural histories. Their thin, moist skins are naturally permeable to water, but are equally permeable to many waterborne pollutants. Most amphibians are also not very mobile, and many live their entire lives within a few square meters, making them especially vulnerable to habitat destruction. Some species have such specific habitat requirements that their entire populations occur in a single spring system occupying only a few square kilometers. The aquatic eggs of many species may be vulnerable to increased amounts of ultraviolet light reaching the earth's surface because of atmospheric changes such as ozone depletion. Pathogens have also been linked to amphibian declines. In particular, the fungal disease chytridiomycosis (caused by the fungus *Batracochytrium dendrobatidis*) has been identified as the cause of declines or extinctions of some amphibian populations in North America. (Although this disease is present in Texas amphibian populations, it has not yet been linked to declines in any Texas species.)

While all the reasons for the rapid disappearance of many amphibians may not be clear, species that were common just a few decades ago are now difficult to find, and some have completely disappeared. Many scientists predict that about one-third of the world's frogs will be extinct or nearly so within a few decades. We should think of amphibians as the "canary in a coal mine" for the environment (miners once carried canaries or other small caged birds into mines because the birds would exhibit respiratory distress or die more quickly than humans when the air in the mine was unsafe). Our amphibian "canaries" are sensitive to changes in water and environmental quality, and their decline is an indicator that all is not right with our environment. In the long term, healthy human populations are also dependent on the quality of the environment, and we should observe the warning signals represented by declining amphibian populations. Correcting the causes of these declines will help not only amphibian populations but human populations as well—ultimately, all living things, including humans and amphibians, depend upon a healthy environment and water supply.

We are fortunate to have a rich assemblage of amphibians in

North America. Altogether there are almost 300 species of amphibians north of the Mexican border, with several areas particularly diverse in salamander species. Because of the wide variety of habitats in Texas, the state is home to a good selection of amphibian species. This book is designed to assist the amphibian observer to both identify amphibians that have been observed and to find habitats in which to observe specific amphibian species. The information presented here may help promote the public's interest in amphibians, which, in turn, may lead to the environmental protection required for Texas amphibians to continue to exist. If we have done our jobs correctly, interest in amphibians by both professionals and amateurs will be fostered. We feel that the encouragement of these interests is an important aspect of ensuring the survival of amphibian species now in jeopardy because of habitat destruction, pollution, and other factors. People tend to protect only those things they are interested in; therefore, furthering interest in these animals should promote their conservation.

NATURAL HISTORY
Seasonality
Most amphibian activity occurs at night, when the relative humidity is higher and the threat of desiccation is reduced. Seasonal activity periods vary from species to species, but terrestrial species are generally least active during hot, dry periods and most active during cool, damp periods. Amphibian activity is generally curtailed when temperatures drop to near or below freezing, although some activity may continue in water below an insulating layer of ice. In general, activity for most amphibian species peaks in the winter or spring, with some notable exceptions. For example, desert-adapted species have activity periods associated with and stimulated by the onset of monsoonal summer rains. One might expect aquatic varieties to be active year-round, but these species are also affected by the seasonality of water flow and changes in water temperature, and are often less active in summer as water temperature increases and stream flows decline. Perhaps the only species in Texas that are unaffected by the seasons are the cave and aquifer specialist species such as the Texas Blind Salamander, since the waters in the aquifer remain a constant 22°C (72°F). But the natural history of

these species remains largely unknown, and their activity may vary seasonally in unknown ways.

Periods of rainy weather often induce an increase in amphibian activity, particularly of terrestrial species. This effect is particularly noticeable in desert and semidesert areas. Amphibian residents of areas far away from permanent water sources rarely leave the area during normal dry periods, but this can change with rain. A dry region normally devoid of amphibians can suddenly have an abundance of salamanders and frogs immediately after a rain, since they will use ephemeral water sources for breeding activity. Many species of amphibians in these areas participate in what is known as "explosive breeding," in which adults emerge from underground burrows and engage in a frenzy of courtship, breeding, and egg laying over a matter of days. Only after breeding do the adults of these species feed, and then return to their burrows with the onset of drying conditions. Meanwhile, for some species, the fertilized eggs hatch, develop into fast-growing larva, and transform into immature copies of their parents all within a two-week period. The juveniles feed until conditions become too harsh, and then, like the adults, they go underground and wait for more rain.

Activity in forest regions is more prolonged than in drier regions: these environments have much higher annual rainfall, and the forest canopy and leaf litter retain moisture far longer than arid regions. But even in forests, rainy periods result in increased amphibian breeding and activity.

The seasonality of amphibian activity also varies by order. Many salamanders tend to be most active in late fall, winter, and early spring, becoming completely dormant during the warm summer months. While some salamander species may occur in high numbers at a particular location, they generally are not encountered as often as anurans; they are more secretive (often living partially subterranean lives), and unlike frogs, they do not advertise their presence with breeding vocalizations.

Frog activity generally peaks in warmer weather than that preferred by most species of salamanders, with peak activity for most species occurring in late spring. Seasonal rains stimulate foraging and breeding activity, and many species will actively forage on humid nights. For example, toads will congregate around streetlights to feed, and tree frogs will assemble around lighted windows to feast on bugs drawn to the lights. Some frog species are ambush predators that lie in wait near the water's edge, where they can be found day or night. Although these species can be found throughout the year, they typically retreat to shaded areas to escape the heat of the summer sun.

Habitat

Amphibians are not the dominant vertebrates in any habitat type in Texas, though they occur in all of them. Their diversity is highest in the more forested eastern regions of the state. Not coincidentally, these areas have the highest rainfall in the state. Most amphibians live near water, and are frequently observed at the margins of bodies of water, where they can take advantage of being amphibious. Amphibians can be found near permanent bodies of water, such as lakes, as well as near temporary pools, such as irrigation flows, seepages, streams, creeks, and swamps. Many amphibians can bury themselves in soil, and some can be routinely found several centimeters below the surface, whereas others live in leaf detritus, in animal burrows, under debris such as rocks or logs, or in trees.

Reproduction

The drive to reproduce is inherent in all animals, since reproduction advances an organism's genes into the next generation.

Northern Cricket Frog (*Acris crepitans*) showing internal vocal sac.

Southern Leopard Frog (*Lithobates sphenocephalus*) showing paired vocal sacs.

Fowler's Toad (*Anaxyrus fowleri*) showing round vocal sac.

Differences in reproductive timing and behavior, particularly the vocalizations of frogs, provide clues to assist in the identification of amphibians in the field. It is often important for identification purposes to note features of reproductive biology such as sex, advertisement calls, or the shape of reproductive structures. It may also be important for the observer to note weather conditions, time of year, and temperature; for example, recog-

Texas Toad (*Anaxyrus speciosus*) showing oval vocal sac.

Southern Leopard Frog (*Lithobates sphenocephalus*) showing axillary amplexus.

nizing these factors can allow one to rule out summer-breeding species when encountering a calling frog in January.

SEX DETERMINATION In amphibians, two basic methods are used in the determination of an individual's sex. First, one can examine primary differences, for example, the presence or absence of particular sex organs. Unfortunately, among Texas amphibians,

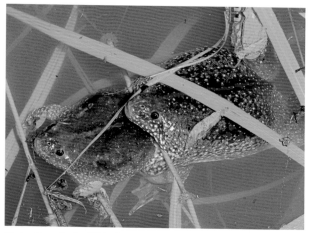

Mexican Burrowing Toad (*Rhinophrynus dorsalis*) showing inguinal amplexus.

only salamanders show notable differences in the shape of the vent or the common opening to the reproductive, digestive, and excretory systems. Next, one can examine secondary differences, such as hormone-dependent differences in body shape or size, color, and behavior. The secondary differences, which are normally linked to courtship, are more obvious in sexually mature individuals, especially during the breeding season.

Simplified characteristics for sex determination in mature amphibians can be seen in the following table.

Characteristics for sex determination in amphibians

GROUP	MALES	FEMALES
Salamanders	Enlarged cloacal margins. Crests, bright colors, and tail filaments in some species during breeding season.	More rotund body shape.
Frogs	Usually nuptial pads in breeding season, and vocal sacs (wrinkled or gray throat).	More rotund body shape.

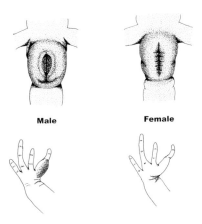

Figure 1. *Top*, Examples of male and female cloaca margins in salamanders and newts. *Bottom*, Examples of frog forearms showing nuptial pads.

EGGS AND LARVAE Identifying amphibians through inspecting eggs and larvae can be a daunting challenge. Amphibian eggs tend to be similar across orders, often with only slight differences in their appearance, making it difficult to distinguish between salamander and frog eggs. The presence of an egg mass can be a guide to the species inside it; likewise, eggs laid otherwise than in a mass can point to particular species. The color, size, and placement of eggs or egg masses are also important indicators. For example, toads in the family Bufonidae lay their eggs in long strands, while frogs in the family Ranidae lay their eggs in large masses. Identification of larva, while still challenging, is a bit easier than trying to match eggs to species. For example, larvae with external gills and appendages are more likely to be salamanders, and those without are more likely to be frogs.

The close examination of gelatinous egg envelopes requires magnification. Although a hand lens can be used, examination is best accomplished with a dissecting microscope, because it provides a stereoscopic view under bright light. Remember to be careful: amphibian eggs are sensitive to drying, quickly desiccate out of water, and may die if overheated.

Salamander larva should be examined carefully, since they are typically very fragile and susceptible to damage from col-

Egg mass of Small-mouthed Salamander (*Ambystoma texanum*).

Couch's Spadefoot (*Scaphiopus couchii*) with egg mass.

lecting and transport. Check the size and shape of the caudal and dorsal fins, length and appearance of the external gills, and overall body size. Color and pattern can also be important for distinguishing the larvae of some species. It is also important to note that several Texas salamander species never trans-

Salamander larvae.

form into a terrestrial adult stage, retaining larval characteristics throughout life.

The larvae of frogs are commonly known as tadpoles. Identification of these amphibians presents a special challenge, since individuals of a single species may appear quite different at different stages of their development. Again, clues to the identification of these animals begin with an examination of the size and shape of the caudal and dorsal fins, although precise identification of species may require examination of the mouthparts. Tooth rows are expressed as a fraction (for example, ⅔, ¾, etc.). The numerator is the number of rows in the upper lip; the denominator, the number of rows in the lower lip. Hyphenated numbers (2–3/2–4) indicate variation in the number of rows. To prevent drying, examine specimens while immersing the larvae in shallow water. Color patterns may also be enhanced in this way.

The eggs of most frogs do not have a tough outer membrane, because they are laid in water and gelatin is the main component of the egg membrane. Tadpoles of most frog species break out of the egg capsule by secreting chemicals from their snout and neck that dissolve through the gelatin.

Although most of the species found in Texas use the normal route of amphibian reproduction via the laying of aquatic eggs, some species use other means of reproduction. Members of the families Eleutherodactylidae (*Syrrhophus*), Craugastoridae (*Craugastor*), and Plethodontidae (*Plethodon*) that occur in Texas lay terrestrial eggs that undergo direct development and hatch as a small version of the adult. The terrestrial eggs differ

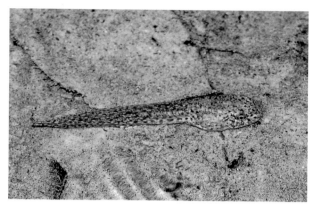

Tadpole.

Froglet still showing tail.

from those of reptiles, however, in that they must be laid in very moist environments (being highly susceptible to desiccation) and lack the complex internal system of membranes present in the reptilian amniotic egg.

Loss and Regeneration of Body Parts

Sacrificing a body part seems catastrophic, but we might all agree that surviving attempted predation is better than death.

Aberrant tail regeneration.

Some animals have evolved strategies to attract a predator's attention to expendable body parts in order to allow the prey animal to escape. Some amphibians and reptiles have the ability to recover from the loss of limbs or the infliction of massive wounds. But only salamanders have the ability to regenerate digits or portions of a lost limb, and only salamanders (mainly plethodontids) and some lizards are capable of regenerating their tails. "Caudal autonomy," the term for the loss of a salamander's (or lizard's) tail, is a very specialized behavior that is induced by an attack.

Glands, Toxins, and Chemical Defenses

All amphibians contain toxins in their skin secretions. In most species, the amount of toxic secretions is inconsequential, and some anurans are regarded as an excellent source of food by humans. In other amphibians, skin secretions can be fatal to humans and other predators.

Amphibians possess several types of epidermal glands, the most important being mucous glands, which secrete mucus to maintain a protective moist film over the skin, and granular glands, which secrete toxins, including peptides and alkaloids. These glands are present in all postmetamorphic amphibians and occur throughout the head, body, and limbs. The glands containing granular cells (for example, parotoid glands) are usually concentrated on the head and shoulder in most species, and are conspicuously noticeable in the true toads and some salamanders. In many salamanders, these glands may also be concentrated on the tail.

In many amphibians, the granular glands may be highly concentrated and especially well developed, particularly in largely terrestrial species. The warts of toads and the parotoid glands found on the back of the neck in toads and some salamanders are actually masses of poison glands. The warts present on toads are therefore natural defensive structures, whereas warts in humans are the result of a viral infection transmitted from person to person—one cannot get warts from a toad! The secretions range in intensity from irritating or mildly distasteful to lethal. The glands occur in the body parts most likely to be contacted by predators; thus, many amphibians will present the head or tail to a predator to ensure that it encounters the noxious secretions. Toads in the family Bufonidae and ambystomatid salamanders have large parotoid glands and stand off against their predators face-to-face, exposing them to the glands' toxins. Others, such as plethodontid salamanders, lash out with their tails, where there are heavy concentrations of these glands. In all cases, it is difficult for predators to grab amphibian prey without being exposed to granular-gland secretions. Even less toxic species, such as tree frogs, still have glandular concentrations that, when taken into the mouth of a predator, secrete a dose of mucus irritating enough to sometimes convince the predator to release them. When handling any amphibian, it is important to remember to always wash one's hands before rubbing eyes, eating, or other activities that involve using the hands, in order to avoid the effects of these secretions.

CONSERVATION

As noted previously, for a variety of reasons, amphibians are especially susceptible to population declines and extinction. In particular, their permeable skins make them susceptible to waterborne pollutants at some or all stages of their life cycles. Amphibians are small organisms, with small home ranges, and when these home ranges are destroyed, most amphibians are incapable of relocating to a new habitat; and at any rate, most adjacent habitats are already occupied by other amphibians. Furthermore, amphibians are susceptible to drying out, and habitat alterations that at first glance seem innocuous may prove disastrous if such disturbances reduce or eliminate the moist refugia upon which these amphibians depend. The introduction of "exotic" species into areas outside their native ranges has been implicated in the decline of many amphibian species. And finally, the widespread introduction of chytrid fungus affects many amphibian populations around the world. With all these threats, why should we have any hope that they might survive? Yet some success stories have occurred. For example, the reclamation and restoration of spring-flow habitats has resulted in dramatic population growth in recent years for the endangered Barton Springs Salamander within the city of Austin.

In many cases, simply listing a species as endangered or threatened is insufficient to stem the tide of its decline. For example, the Houston Toad has been listed as a federally endangered species since 1970. Yet today, this species occurs in only a tiny fraction of the area it occupied when it was listed. Only in recent years have aspects of its biology become well enough understood for herpetologists to be able to identify critical aspects of its habitat: intact post-oak woodland savanna or pine woods in areas near appropriate-depth breeding pools within areas of deep sandy soils. Unfortunately, this specific type of habitat is declining (not surprisingly) as fast as the toad is. In fact, the decline of appropriate habitat is causing the decline of the toad. Only with increased knowledge gained from research and a caring and educated public will this species stand a chance of surviving this century.

We see the effects of habitat loss every day. Houston has a booming economy and is converting forest, prairie, and wetland habitats into homes and commercial properties almost faster than the city can record the changes. Austin has shopping, business, and population centers being constructed at a dizzying rate over environmentally sensitive aquifers. Ponds and streams are being filled in at an alarming rate, and species that were once quite common are disappearing from the countryside, possibly being removed from our lives forever. Perhaps these economic changes and growth are necessary, both for the good of the country and for the people living in it. But the price of economic growth need not be the loss of environmental health. We should be smart enough and, hopefully, caring enough to find ways to sustain economic growth while protecting critical habitats for amphibians and other creatures.

Unfortunately, even the existence of pristine habitat no longer guarantees that amphibians will survive. Amphibian populations have been monitored since the 1980s in several parts of the world, and in 1991 the Declining Amphibian Population Task Force (which was merged into the International Union for Conservation of Nature/Species Survival Commission [IUCN/SSC] Amphibian Specialist Group in 2006) was established to coordinate research throughout the world. When this book went to press, IUCN/SSC findings from 2007 suggested that 36 amphibian species are believed to have gone extinct worldwide since 1980, and 3 of those disappeared before they were even formally named. In addition, another 26 species that have not been seen for 5 years or more may now be extinct, and another 91 are critically endangered (including 3 Texas species). Many species, especially the secretive ones, could go extinct before we know of their existence. Small amphibians are often very delicate and unable to adjust well to disturbance of their habitats.

Many factors contribute to amphibian declines, some of which are listed below.

Habitat Destruction

The wetlands upon which most amphibians depend have long been a decimated or neglected habitat. Often regarded as "useless" at best or a "nuisance" or "health hazard" at worst, wetlands have been systematically targeted for destruction for hundreds of years. Wetland soils have long been recognized as useful for agriculture when drained, and this has been the fate of the vast majority of seasonally flooded wetlands throughout the United States and the world. The U.S. government sponsored this natural habitat destruction: it encouraged wetland reclamation projects from 1950 through the 1970s, with little thought for native fauna and flora, especially amphibians. Reclaimed land was used for forestry, agriculture, residential housing, or industry. Even less thought was given to the ecological importance of wetlands. Fortunately, since the 1980s, many people have recognized that in addition to providing homes for diverse communities of wildlife, wetlands slow runoff, store and filter water supplies, and reduce the effects of flooding and erosion. Unfortunately, this new awareness of the importance of wetlands has come too late for many areas, where the destruction of these habitats has already taken place.

Pollution

With increases in human population comes a proportional (or even exponential) increase in pollutants. Chemical pollutants increase in the air, on the ground, and in the water because of automobiles and public transportation, freight carriers such as trucks and trains, mining, refining and chemical processing, residential chemical and pesticide use, agriculture and insecticides, water and sewage treatment, and a wide variety of other sources associated with human activity and industrial society.

Unfortunately, many of these chemical pollutants are a necessary part of an industrial society. The more obvious include plastics and emissions from automobiles, the less obvious include insecticides and herbicides. Insecticides are a prime example of an often-necessary pollutant having both positive and negative consequences. These chemicals kill problem insects, such as mosquitoes, which are known to be carriers of devastating diseases such as equine encephalitis, yellow fever, West Nile virus, malaria, and dengue fever. Without insecticides, these diseases would likely occur more frequently. Yet pesticides are also harmful to many organisms, including most amphibians. Continued research and improvements in pesticide specificity and management are necessary for the future.

Introduced Species

The introduction of nonnative species into new areas usually has a negative effect on native wildlife, and amphibians often bear the brunt of this. The introduction of nonnative species is not restricted to "foreign" species imported from other countries; it also refers to native species being moved from one location to another. One of the best examples is the American Bullfrog (*Lithobates catesbeianus*). The American Bullfrog is native to the eastern and central parts of the United States and Canada. Its natural diet includes a huge variety of prey, both invertebrate and vertebrate, including other species of amphibians and reptiles. The American Bullfrog has been widely introduced outside of its range and is now a problematic invasive exotic in the western United States. This large frog does quite nicely in appropriate new habitats, often quickly adjusting its dietary habits and consuming local fauna, including endangered species. In some areas, it has displaced (or consumed) populations of local native frogs. The introduction of sunfish into naturally fishless ponds has destroyed the breeding habitats of many salamander species. In this case and others, simply moving a locally abundant fish from one pond to another just a few meters away can result in the extinction of a local amphibian population. Perhaps the most common invasive exotic is one that is often overlooked—our pets. Dogs and, particularly, cats have been well documented to take a toll on native wildlife when allowed to roam freely or become feral.

The introduction of nonnative species can take its toll in other ways. For example, salamander larvae, such as those of Tiger Salamanders, are regularly sold as live fish bait, and many of these bait salamanders originate far outside Texas. Unused bait salamanders discarded at the end of a fishing trip can survive to reproduce with native populations of a similar species. The result is the "pollution" of the genetics of these local populations with exotic genes. This interbreeding may have unforeseen effects on the local population, which is adapted for the specific environmental conditions in which it is found, while the exotics are adapted to a suite of different environmental pressures. If the local population becomes too "polluted" with exotic genes, it can lose genetic fitness and the ability to survive.

Collecting and Trade of Animals as a Cause of Decline

Although the collecting of amphibians from the wild has been implicated in the declines of some species, the effect of collecting on most species of amphibians is completely unknown. Perhaps the most well-known case of collecting having reduced local populations of amphibians in the United States is the collection of leopard frogs (*Lithobates pipiens* complex) for preservation and sale to high school biology classes for the purposes of dissection. Only a few common native amphibians are collected commercially for the pet trade (for example, Green Treefrogs [*Hyla cinerea*] are commonly sold at pet stores), and although there is concern that some rare species may be targeted for collection, the effects or potential effects of these collections remain largely speculative.

Chytridiomycosis

Chytridiomycosis, an infectious disease specific to amphibians, is caused by a parasitic fungal pathogen named *Batrachochytrium dendrobatidis* (Bd). Bd infects the keratin structures within the skin of adult amphibians and the mouthparts of larval amphibians, but only in species susceptible to the infection. Some species carry Bd but are not affected by it. Bd has been identified on all continents where amphibians occur and has been implicated in the decline of more than 250 amphibian species worldwide. Most of these declines have occurred in mountainous regions of Australia and the Americas. It is prevalent in mountainous regions because the fungus is most virulent between 13° and 23°C (59° and 73°F), and the virulence significantly drops off at temperatures above 27°C (81°F). Luckily for Texas amphibians, temperatures throughout the state regularly exceed the optimal temperature range for Bd growth for long periods each summer. That said, Bd has been found in several species of native Texas amphibians; however, no declines in Texas have been attributed to it.

Even though Bd has not become a problem in Texas, people working with amphibians should take precautions when searching in aquatic habitats. Bd can be spread between sites on field gear like dip nets and boots. To avoid the spread of Bd, one should soak all field gear used when searching for amphibians in a 10 percent bleach solution before using the same equipment at another locality. Also, Bd is thought to be spread by transporting live amphibians across borders or by releasing animals into different populations. Therefore, amphibians should never be caught in one location and released into another.

OBSERVING AND COLLECTING AMPHIBIANS

Observing and collecting amphibians in the wild can be an enriching and rewarding activity, and one that brings people closer to nature. This field guide is written for those who want to go into the field, experience where amphibians live, and observe them in their natural habitats. Although many people will choose to follow the National Park Service slogan "take only pictures," others may choose to collect specimens for personal observation in a captive setting. Captive husbandry of these fascinating animals is often quite rewarding and can provide insights unobtainable in any other way. Collecting for such a purpose should not be discouraged, provided it is done responsibly and in moderation. Furthermore, collecting for the purposes of providing voucher specimens for academic museum collections is a vital way to further our understanding of temporal and spatial changes to amphibian species and populations. Museum voucher specimens can also be used in a variety of research programs, including the investigation of amphibian evolutionary relationships and the documenting of historical distributions of species in the face of changing climatic conditions.

Observing Amphibians

To observe amphibians, one must go where they live. Each species has its own unique behaviors and habitat requirements, and a prospective observer must learn those behaviors and requirements. Tips for observing particular species may be gleaned from the species accounts included later in the book. But some generalities are presented here.

Most terrestrial amphibians are active at night and hide by day. This presents two opportunities for observation. To observe these species in their active cycles, one must use a bright light, either a handheld flashlight or headlamp, and search in their habitats. Move slowly and carefully, panning the light around at ground level or, for arboreal frogs, in the lower branches of trees. While doing this, the observer should look for the amphibian itself or for "eyeshine" reflected off its eyes. In particular, scan the banks along bodies of water, since members of many amphibian species will sit along the shore in wait for their prey. To observe these same amphibians during the day (when they are hiding) involves searching for cover objects, including boards, rocks, logs, or piles of leaf litter. Carefully lift up the object and look underneath for resting amphibians. The cover object should be just as carefully returned to its original position, whether an amphibian is present underneath it or not. If a target amphibian is present under a cover object, it typically can be caught by hand for close observation and then released. Do not lift rocks or logs so large that they cannot be put back easily, and do not destroy any cover objects. In particular, logs should be left intact on the forest floor, since they provide homes for a wide variety of animal species, including amphibians, and the destruction of these refugia can have a detrimental effect on the populations of the animals that one is attempting to observe. Recent research has indicated that rocks moved as few as 30 mm (slightly more than an inch) from their original position will harbor fewer reptiles than undisturbed rocks (Pike et al. 2010).

Frogs are most easily observed when they are in breeding choruses. These may often be heard for miles as male frogs advertise their readiness to mate. A chorus of frogs is usually easily located by following the calls to a pond. Upon the arrival of a human observer, many frogs will cease calling. But if an observer turns off his or her light and waits quietly and patiently without moving, the chorus will typically resume calling. Locating a single calling frog may prove more difficult, since many species are small and call from within the cover of grass or brush. Having more than one observer really helps in this case, since two people can often "triangulate" the calling frog by standing a few meters apart and shining their lights toward the source of the call. Generally speaking, where the lights cross will be the approximate location of the calling male. Finally, some frog calls can be successfully mimicked by the human voice. Doing so may result in any frogs present starting to call in response to the mimicked call.

Observing aquatic amphibians presents special challenges. With the use of a light, spring-dwelling species can often be observed at night as they walk about on the substrate in their springs. They may also be found during the day under small stones at the water's edge. But when exposed to light, these secretive animals may make a mad dash for cover, leaving the searcher with only a glimpse of a fast-moving salamander. Most easily caught by using a small aquarium net, they can be observed briefly by placing them in water in a small container. The spring-dwelling salamanders native to Texas are so small and fragile that they can be especially difficult to maintain in captivity, and their captive maintenance is best left to keepers with extensive experience with these delicate creatures.

Aquatic salamanders dwelling in the sluggish streams of

eastern Texas pose different challenges. Amphiumas, sirens, and mudpuppies all inhabit largely murky waters, where they frequent logjams and leaf-litter piles. They may occasionally be viewed at night as they swim about in the water (amphiumas tend to not react to light, while sirens tend to vigorously avoid it), but it is a lucky amphibian observer indeed who makes such an observation. These amphibians may also be searched for by pulling large, long-handled nets through the water. Once a specific habitat that contains these species is located, they may be found to be surprisingly common.

To enhance one's knowledge of amphibians, it is often helpful to maintain a journal in which useful information on observations is maintained, including location, date, time, temperature, moon phase, species, and number of individuals observed. This information can be consulted in the future, allowing the observer to rely on documented data rather than a faulty memory. Furthermore, the information contained in such journals can be used to assist researchers to understand and target local amphibian populations.

Photographing Amphibians

Most of the photographs in this book were taken with a single-lens reflex (SLR) camera with a macro lens and an accessory flash or flashes. Those images taken before 2006 were shot with film cameras, but all the recent photos were taken with digital SLRs. Focal length for the macro lenses ranged from 90 to 105 mm. The use of flash allows images to be taken at a high f-stop, resulting in increased depth of field. Some of the posed photos were taken with two separate off-camera flashes mounted with diffusers, while most of the photos of calling frogs and the in situ photographs were taken with a single flash—either off camera or even with the pop-up flash integral to the camera.

When photographing an amphibian that has been collected, the first consideration for a quality photograph is a suitable background. Moss beds, leaves, and other relatively simple, nonreflective surfaces allow the animal to stand out. Frogs may be posed in such a way that their faces point slightly toward the camera. Elongate amphibians such as salamanders should be posed with their tails coiled either toward or away from the

Example of photography setup.

camera—otherwise, details of the animal's body cannot be seen clearly.

A clean glass aquarium may be necessary for the photographing of aquatic species. Advance preparation of an aquarium with clear, clean water over a substrate of sand that has been allowed to settle on the bottom will be necessary to get quality photos. Put the salamander or tadpole in the water and allow it to acclimate. Finally, place the flash at the top of the aquarium so that it will shoot down into the water and simulate sunlight, and then photograph through the side of the aquarium.

It is a greater challenge to photograph wild amphibians in situ. Calling frogs are perhaps the easiest to photograph, since they are focused on reproduction; if approached slowly and carefully, they can be photographed. Wild frogs that are not calling are most easily approached at night while they lie in wait for prey. The same frogs may be far more wary, and virtually unapproachable, during the day. Terrestrial salamanders are rarely seen on the surface except on the rainiest of nights—conditions that are usually unsuitable for photography. But spring-dwelling species can often be photographed quite easily as they walk about on the bottoms of the clear pools they inhabit.

Collecting Amphibians

Although collecting animals from the wild is becoming more and more controversial, we feel strongly that the decision to collect should be a personal one (provided the collecting is limited and is done legally and responsibly). Many people find collecting for a personal captive collection to be more rewarding than purchasing an amphibian from a store, because maintaining an animal that has been caught in its natural environment often brings back the memories of "the chase." It is important to observe local, state, and federal laws when collecting in the wild. In Texas, a collector must have either a permit issued by the state or federal government or a Texas hunting license to collect nongame species. Even with these permits, it is important to recognize that certain laws are strictly enforced and some animals cannot be captured without special state and federal permits. The reader should become familiar with requirements by contacting the Texas Parks and Wildlife Department (TPWD; www.tpwd.state.tx.us) for more information.

A final note on collecting: the advantage of collecting amphibians is that one group or another can be collected year-round. Unlike snakes, which tend to stop moving during winter, some amphibians can be found throughout the year in Texas.

For our university work, we collect voucher specimens for proof of location or season. We make these collections sparingly, and we record and archive complete scientific information on each specimen. Usually, only one or two voucher specimens for a particular study area, season, or objective are required. For one ongoing project, we collect only two voucher specimens per species every 5 years. For many museum-collection purposes, we attempt to find what are referred to as DOR (dead on the road) animals so that we will not have to kill one. Although this strategy is more applicable to reptiles than amphibians, fresh DOR specimens are often in good enough shape to be prepared for museum storage. Sometimes a vouchered animal is kept alive by a collector or in a museum collection until its natural expiration, when it is prepared and stored permanently in a museum collection. Although photographs may be vouchered, in most cases a specimen is preferred, since a specimen provides much more data than a photograph. Today, it is increasingly important to take tissue samples for DNA work—such tissues must be taken before fixation in preservation media. In all cases, a collector should keep copious field notes and make them available, whenever possible, to local "herp" groups for publication in newsletters when something new or noteworthy appears.

Legal Aspects of Collecting Amphibians

A number of recently enacted laws in Texas apply to the collecting and trade of native nongame species, including amphibians. These laws apply also to the importation of the same species. In addition to state laws, there are local ordinances that govern the taking and maintenance of certain amphibian species. These laws, which vary according to locality, are in a continual state of flux. It is important to know the city, county, state, and federal laws pertaining to the area where a person is collecting.

The legal aspects of collecting and maintaining animals become more complex when the subject involves importing amphibians, purchasing imported animals, moving animals from place to place, or simply maintaining animals in captivity. We do not deal with legal issues regarding the importing of foreign and exotic animals, since they really do not have a place in a field guide for Texas. But as mentioned earlier, introduced species can displace native species in their natural habitats if released or if they accidentally escape from a cage into the environment. We do not condone the purposeful introduction of any nonnative species into Texas habitats.

Legislation and regulations have been established to provide protection and guidelines for collecting, dealing in, and maintaining reptiles and amphibians. Some of these laws have been written for the protection of rare and endangered species, some for the protection of habitats, and some to protect the keeper and the public. For anyone planning to collect, it is vital to set limited objectives and know the legal consequences before attempting to collect, import, or keep any reptiles or amphibians.

There are several relevant levels of legal jurisdiction for the collecting and keeping of amphibians. Ignoring any of these may result in a citation, a fine (if one is found guilty), or, in cases involving threatened and endangered species, jail. These levels of jurisdiction include the following.

LOCAL LAWS These are laws passed by cities and counties. Most of these governments have ordinances concerning collecting or keeping amphibians. Many ban the keeping of certain types of reptiles (in particular, large or venomous snakes or invasive exotics) in personal homes. Check with the local governments before collecting or keeping amphibians. The local humane society may be the logical place to begin. Follow this with a phone call to a local animal control officer.

TEXAS STATE LAWS At present in the state of Texas, a hunting license is required to collect most amphibians. Most species can be collected only for noncommercial purposes and in limited numbers (fewer than 6). Only a few species can be collected in excess of this figure, and then only with a commercial collecting permit. Further, only a limited number of amphibian species can be collected for commercial purposes, and in all cases, a commercial collecting permit of some sort is required. Contact the TPWD for additional details.

FEDERAL LAWS Aside from regulating the collection of federally threatened and endangered amphibian species, the U.S. government allows the states to regulate their own native species. But when a violation of state laws involves the transport of animals across state lines, federal law has been broken (the Lacey Act), and federal law enforcement agencies can take jurisdiction.

The Lacey Act, which is administered by the Department of Agriculture, states that the acquisition of any plant or animal in violation of any sovereign state's law is a violation of federal law. In addition, the "injurious wildlife" provisions of the Lacey Act concern the importation and transport of such animals as mongooses, walking catfish, venomous snakes, and toxic amphibians; these provisions are administered by the U.S. Fish and Wildlife Service (in the Department of the Interior).

The Endangered Species Act is designed to protect rare and endangered animals. The U.S. Fish and Wildlife Service can provide a listing of currently recognized endangered amphibians and reptiles.

Permits and Collecting Amphibians in Texas

The permitting and legal issues regarding the hunting and collecting of amphibians, as with other game and nongame animals, are always in a state of evolution, so we recommend that the reader contact the TPWD for current information. The following guidelines are taken from a recent *Texas Parks and Wildlife Outdoor Annual: Hunting and Fishing Regulations* (2009–2010):

- Amphibians are considered to be nongame animals, with strictly aquatic amphibians considered nongame fish.
- A hunting license is required of any person, regardless of age, who hunts any animal or bird in this state (Texas).
- A hunting license is required to take non-protected turtles and frogs.

Nongame species in Texas have been further regulated by the quantity that can be kept in captivity. Species not listed as threatened or endangered have been put on either the "White List" or the "Black List." Species on the Black List may not be sold and can be kept only in quantities of 6 or fewer. Species on the White List may be sold, and 25 or fewer may be kept without a nongame dealer permit. The permit is needed when maintaining more than 25 White Listed animals or when selling these animals.

In 2011, the Texas State Legislature adopted legislation legalizing the collection of amphibians and reptiles from public right-of-ways (reversing legislation passed in 2007). This legislation requires amphibian collectors to wear reflective vests and to park safely off the roadway, and prohibits the use of artificial lights from a vehicle. This legislation also directs the Texas Parks and Wildlife Department to create a "herp stamp" endorsement, which a prospective amphibian collector must purchase along with a hunting license. On September 1, 2011, this stamp was made available. Although it is not illegal to search for amphibians for legitimate photographic purposes, be advised that an amphibian photographer is likely to be suspected of collecting. When using a light to photograph or search for amphibians at night, consider making a courtesy telephone call to the local game warden.

In addition, we recommend that readers refer to current regulations regarding the status of resident and nonresident collectors, as well as the general hunting and fishing regulations that pertain to all wildlife users. For further questions or for more information, contact the TPWD Law Enforcement Office at 1-800-792-1112 or 512-389-4800, Monday through Friday, 8:00 a.m. to 5:00 p.m.

Threatened, Endangered, or Protected Nongame Species

It is unlawful for any person to hunt or collect threatened, endangered, or protected nongame species. Proper documentation must accompany goods, whether purchased or sold, made from threatened or endangered species. For a complete current list of threatened and endangered species, and the regulations relating to breeding threatened or endangered species, contact the TPWD. There may be other, more encompassing federal or CITES (Convention on International Trade in Endangered Species of Wild Fauna and Flora) regulations, and any amphibian hunter or collector should be aware of them.

MAINTENANCE OF AMPHIBIANS
Maintaining Amphibians in Captivity

There are several reasons to maintain amphibians in captivity: they make fascinating pets, they require relatively small spaces, and some have maintenance requirements that are easily met in small homes, apartments, or offices. Others may be kept (by professionals) for scientific study or for conservation efforts. But it must be noted that amphibians in general tend to be more difficult to maintain in captivity than similar-sized reptiles, because they are much more sensitive to captive environmental conditions.

It is important to remember that only a relatively small number of amphibian species are suitable for home vivarium culture. Many species, particularly some of the smaller salamanders, feed upon invertebrate prey that are too small to be readily available to the hobbyist. Winter-active species may require temperatures cooler than those normally maintained in a home, and these same species may in nature remain dormant underground in cool refugia during the heat of the summer. Generally speaking, the most suitable amphibian captives include the larger generalist species, which in nature feed upon a wide variety of prey and are active at a wide range of temperatures.

If the collector is taking animals for keeping at home, it is important to become familiar with the various diseases and parasites that can accompany captured animals. References are available that give excellent presentations of the causes and cures of these problems.

Because they respire across their delicate skin, amphibians take up any toxins in the local atmosphere and environment. Their cages or enclosures must be spotless, and the water in their environment must be absolutely free of chemical pollutants. Chemical contaminants such as chlorine in water, disinfectants and other cleaning material, air fresheners, or smoke from a fireplace or candles can affect the survival of amphibians in captivity.

Some of the most suitable amphibians for captive husbandry are those that live in aquatic environments; their care differs little from that of freshwater fish. For example, amphiumas and sirens can do quite well for years in a filtered aquarium with a secure lid, as long as they are provided with an adequate food supply. One difference from freshwater fish is that these species (particularly the large, vertebrate-feeding amphiumas) tend to produce large quantities of waste, meaning their tanks require more frequent filter cleaning than those of typical freshwater fish. Some species may prefer a cage design that is about half water and half soil, while others can be kept in shallow water on a moist substrate of gravel or sphagnum moss. Toads, as well as frogs that spend most of their lives on land, may need water only for a short period for breeding. If a caretaker is not interested in breeding these animals, they can be kept in a terrarium all year round, provided that adequate humidity is maintained in it. Beginners should avoid overloading the environment with flashy features such as running water or numerous living plants until they are comfortable with their ability to maintain the animals.

Most beginners make the mistake of thinking that amphibians are like reptiles and so prefer warm environments. Many amphibians, including those from the tropics, prefer cooler temperatures than most reptiles. Even those found in deserts usually do not emerge until the occurrence of cool weather or cooling rains. Since most amphibians lead nocturnal, secret lives, light should be minimized; they often become stressed if forced to live under the glare of bright lights.

Humidity is important, and even the terrestrial species should have their cages sprayed with water regularly to ensure they do not dry out. Since amphibians in nature often do not do well in stagnant atmospheric conditions, ventilation is very important.

Most amphibians are insectivorous, and even large species, such as the American Bullfrog, can live their entire lives on a diet of insects and other invertebrates, like worms. The correct insects and earthworms can be purchased at a local pet store or from wholesale suppliers. Supplementation of an animal's diet with wild insects may provide important nutrients lacking in store-bought insects; however, one must be aware that insects from the house or from around the yard might have been exposed to chemicals such as pesticides, herbicides, or fertilizers, all of which are harmful to amphibians. Supplement with wild insects only if they are certain to be chemical-free. Larger carnivorous species, such as American Bullfrogs, Three-toed Amphiumas, and Tiger Salamanders, may feed on newborn or young mice. Most tadpoles from frogs are herbivorous and will usually ingest algae built up on the surface of plants and rocks, but they will also eat flaked fish food and a variety of other special diets. Be aware that some of these tadpoles may become carnivorous and attack other tadpoles in their environment. We recommend that the caretaker search for specialized publications dedicated to the maintaining of a specific species.

It is possible to enjoy amphibians without necessarily isolating them in an aquarium or terrarium. A small garden pond with an overgrown area around the edge can provide an ideal habitat for many species of frogs, toads, newts, and salamanders. This method of maintaining amphibians also helps conservation efforts and replaces habitat loss that occurs through urban development. Some amphibians will visit the pond to breed, and then disappear into the surrounding area. Others will remain close to the habitat. Some can be encouraged to remain as local residents if there are hiding places such as logs, rock piles, or other places where they will feel at home. The use of chemical substances (such as weed killers and fertilizers) must be avoided for the pond to be a success. If prepared correctly, such a garden should not require pesticides, since natural insect predators and amphibians residing in the pond will keep pests under control.

We must reiterate that in many places it is illegal to introduce nonlocal species; however, even a small pond that provides a home for a single species of frog or salamander can be very interesting.

Creating a Natural Setting

Rather than collecting amphibians for a vivarium, some people try to attract amphibians to a backyard wildlife habitat. In Texas, toads usually do the best in restricted areas and are often found naturally in residential yards, even in large cities such as Houston, Dallas, San Antonio, or Austin. Not only are these settings ideal for observing amphibians in the wild, they are generally much less trouble to maintain, and definitely cheaper. An amphibian setting or amphibian house designed for the special needs of these creatures can be home for frogs, toads, and salamanders. In East Texas, amphibians are often found in water-meter boxes, and as long as they have a protected, moist environment, they can do well.

There are commercially available "toad houses," but it is easy to make a toad hideout by recycling an old clay flowerpot. Break off a small section of the lip of the pot, and then place it in a shady part of the yard, upside down. The pot helps retain humidity and provides a cool, dark place for toads, frogs, and other moisture-seeking creatures. A shallow water dish can be added to make the abode more attractive to amphibians. We have discovered toads living quite comfortably in environments such as these in arid far West Texas. The advantage to having an amphibian house is that these animals, especially the toads, relish slugs and insects and for this reason make excellent garden inhabitants. But these toad homes are also ideal places for other garden inhabitants to hide, such as snakes and centipedes, so one must always take care when lifting the "house" to prevent accidents.

Handling Amphibians

The first, and only, rule of handling amphibians is not to do it unless necessary. All amphibians should be restrained as little as possible except for a good reason. If it becomes necessary to hold an amphibian, as when collecting, or restraining one for health, or determining sex, it is important to know the correct procedures and methods. Also remember that all amphibians have skin excretions, many of which are toxic.

There are several reasons to avoid handling an amphibian. They are very fragile animals and are liable to be injured if squeezed too tightly in certain parts of their bodies, or when attempting to escape. More importantly, their delicate skins absorb any chemicals that may be present on the handler's hands and arms. In our everyday activities, we can come into contact with hundreds of chemicals. Our skins, which are basically impervious to these chemicals, keep our internal system isolated from solutions, insecticides, oils, etc. But an amphibian's skin readily absorbs all these.

Specimens of the smaller species tend to be the most delicate, and when possible, it is best to coax them into a small, clean plastic container or a polyethylene bag for transportation or examination. Ensure that the container has been thoroughly cleaned and rinsed with clean water. Also remember that most tap water contains chlorine, as a dissolved gas, and it is best to leave the container open for some period of time before using it so that the chlorine can evaporate (24 hours appears to be a safe period).

Specimens that are held in the hand should be grasped firmly. It is important not to hesitate when capturing and holding an amphibian. Many amphibians move quite well when cold, and unlike reptiles, removing their heat source for a short period does not make them more manageable.

AQUATIC AMPHIBIANS Aquatic amphibians are most easily caught in a hand net and should be held only in hands that have been first dipped into water. A clean plastic bag or container (such as a gallon jar or bucket) is useful for controlling them for short periods of time.

FROGS Small frogs and toads may be captured by completely enclosing them in one's hands and gently closing the fingers around them. Larger specimens can be grasped by their hind legs. Very large specimens, like American Bullfrogs, can be picked up with the fingers spread on each side of the neck and head and the front legs gently grasped to each side.

SALAMANDERS These are usually very easily captured and held in the palm of the hand with the index finger beneath the throat and the thumb positioned slightly on the top of the head or neck. Salamanders tend to move relatively slowly on land compared to other amphibians, such as frogs. But streamside-dwelling species may be particularly slippery, and a small dip net may facilitate their capture.

MUSEUM AND PRESERVED AMPHIBIAN SPECIMENS

We do not encourage private individuals to build a collection of preserved amphibian specimens. Such specimens are of much greater value when housed in an active research museum, and should be deposited in one. But public school teachers or nature center directors may find it useful to keep preserved examples of local species for educational purposes. While an individual keeper or collector might preserve deceased captive specimens or those found DOR, we do not have a section devoted to preservation techniques. There are many references that cover these techniques in detail, but we recommend that any reader interested in contributing preserved specimens to a museum or a teaching collection visit a local university for advice from a professional museum curator.

It is important to point out that scientific institutions are not necessarily interested in taking specimens from unknown sources. Generally, a collector must develop a relationship with an institution before it will trust that a specimen was legally collected and the accompanying information is accurate.

SCIENTIFIC AND COMMON NAMES

Taxonomy is the field of biology that categorizes all living things into groups and assigns names to those groups. Originally, this field was primarily concerned with naming organisms in order to facilitate communication between biologists speaking different languages. Organisms and groups of organisms are assigned distinct names at different taxonomic levels, which range from the all-encompassing kingdom down through increasingly narrower levels of classification until the most specific level, the species, is reached. Biology students may remember learning "*King Phillip Came Over For Great Spaghetti*," a mnemonic for kingdom, phylum, class, order, family, genus, and species. In recent years, as DNA and biochemical evidence have allowed a deeper understanding of the relationships between various bacteria and eukaryotic organisms (those whose cells possess a nucleus), the rank of domain has superseded kingdom at the top of the classification hierarchy. Perhaps most importantly, within this taxonomic system, every species of living thing has been assigned a scientific name consisting of a genus name and a specific epithet. This two-part scientific name is a unique combination that identifies a single species.

An example best illustrates how this naming system works. The Green Treefrog, a widespread species found throughout the eastern two-thirds of Texas (as well as much of North America), is classified as follows:

DOMAIN: Eukarya (name for the group that includes eukaryotic organisms, whose cells contain a nucleus)

KINGDOM: Animalia (name for the group that includes all animals)

PHYLUM: Chordata (name for the group of vertebrates and closely related invertebrates)

CLASS: Amphibia (name for the group of four-legged vertebrates with aquatic larvae and terrestrial adults)

ORDER: Anura (name for the group that includes frogs and toads)

FAMILY: Hylidae (name for the group that includes the tree frogs and relatives)

GENUS: *Hyla* (name for a group of tree frog species)

SPECIES: *Hyla cinerea* (specific name for Green Treefrogs)

Notice that the species name includes the genus name. The specific epithet *cinerea* (Latin for "ash-colored," describing a preserved specimen) is insufficient to completely and specifically identify the Green Treefrog scientifically, in much the same way that the name "Robert" is insufficient to specifically identify an individual person; complete identification requires the addition of a last name such as "Smith." While with human names there may be multiple "Robert Smiths," with scientific names there is only one *Hyla cinerea*, and that name can be applied to only a single species.

Also notice that the genus and species names are italicized. This reflects the origin of these words in Latin (in most cases). Latin was the language of scholars and scientists when the scientific classification system was first created, so it was natural for those scientists to use Latin in their naming regimens, especially because their goal was to create names that all scientists could understand, regardless of their native tongues. Today, the system of biological nomenclature continues to use Latin, or Latinized versions of words from languages such as Greek, English, or Spanish, to name organisms. These names are assigned to species using rules outlined by the International Commission on Zoological Nomenclature (ICZN).

While taxonomy itself may seem fairly straightforward, it is generally deemed desirable for the classification and naming of organisms to reflect our understanding of how organisms are related to one another. The study of how organisms are related to one another is known as phylogenetic systematics. The goal of phylogenetic systematics is to use the characteristics of organisms—including morphological, behavioral, and genetic features—to infer the relationships between populations within a species, to infer relationships between different species of related organisms, and to infer relationships between different genera, families, and orders. Whereas the field of taxonomy is governed by a system of rules, phylogenetic systematics often seems quite chaotic to laypersons, since scientists propose conflicting hypotheses regarding the relationships between the groups they study. This is, in fact, quite representative of the scientific process: over time, evidence will mount in favor of one hypothesis over another, and a better understanding of the relationships between organisms will ultimately result.

Unfortunately, the process of phylogenetic systematics does not always lead to stability in taxonomy, and name changes can frequently occur. This process is often trivialized by the categorizing of phylogenetic systematists as either "lumpers" or "splitters." Lumpers are those biologists who combine taxonomic groups under a single name, while splitters divide members of a single taxonomic group into two or more new categories. Perhaps a third group could also be recognized—the "reshufflers"—

namely, those researchers who move named groups from one classification to another.

This is an exciting time to be a phylogenetic systematist; technological advances allow researchers to access larger and larger fractions of organisms' DNA, which, in turn, allows ever-increasing amounts of data to be used to infer relationships between groups of organisms. Consequently, the names of species today are in a state of flux, with new taxonomic arrangements regularly being proposed. This applies to amphibians as well, and species names that naturalists learned a decade ago have, in many cases, been changed to reflect science's new understanding of phylogenetic relationships.

These changes present as much a challenge for a field-guide author as for an amateur naturalist interested in keeping track of the names! To help everyone stay current, the Society for the Study of Amphibians and Reptiles (SSAR) publishes a standardized list of names: *Scientific and Standard English Names of Amphibians and Reptiles of North America North of Mexico, with Comments Regarding Confidence in Our Understanding* (Crother 2008). Because of the nearly constant changes in taxonomy, this list is updated periodically online at http://www.ssarherps.org/pages/comm_names/Index.php. Note that this is only a list of standardized English names, and that the scientific names outlined within it are a reflection of the authors' "confidence in understanding" of the relationships. While there are lists of names published by other sources in North America, the SSAR list is generally considered the most widely accepted one, and so it should be the starting point for anyone wishing to keep track of changes in taxonomy.

For this field guide, the published SSAR list was used as a primary reference, but we deviated from the list in several areas. First, when a new genus names replaced a long-standing, well-known "traditional" name, we included the older names in brackets following the new genus name (for example, *Lithobates [Rana] catesbeianus*, or American Bullfrog). Second, when we felt that insufficient evidence supports recent taxonomic changes, we chose a conservative approach and used the older taxonomic arrangement. Finally, when we felt that the appellation of a name is in error, we presented the taxonomic arrangement—following the rules outlined in the ICZN— that, in our opinion, best reflects the current understanding of species relationships. In particular, we differed from the SSAR list in the following three respects.

1. Use of the generic name *Syrrhophus* rather than *Eleutherodactylus* for the chirping frogs, *S. marnocki*, *S. guttilatus*, and *S. cystignathoides*

   The genus name *Eleutherodactylus* has long been used for a wide variety of frogs in the family Eleutherodactylidae, and has widely been applied to species only distantly related to one another. The name *Syrrhophus* very specifically refers to a group of closely related crevice-dwelling terrestrial frogs that are more closely related to one another than to other members of the family, and so we retain the name *Syrrhophus* for Texas species of chirping frogs.

2. Use of the name *Pseudacris feriarium* instead of *Pseudacris fouquettei* for Texas populations of Upland Chorus Frogs

   The Striped Chorus Frog complex (*Pseudacris triseriata* complex) has been subdivided in recent years into as many as 5 different species, based largely on the use of mitochondrial-DNA evidence. While we feel that there is sufficient evidence for the designating of multiple species within the complex, the species boundaries for all species have not been well delineated. Furthermore, some of the mitochondrial-DNA evidence used to split off species within the complex was unable to distinguish a morphologically distinct species (*Pseudacris clarkii*), suggesting that the DNA fragment used may not be appropriate for making species-level designations within the complex. We could not come to a consensus on which name should be used for this species, and as a compromise position, we use the name *Pseudacris feriarium* for Texas populations of Upland Chorus Frogs.

3. Identity of Lower Rio Grande Valley populations of *Siren*

   Sirens in the Lower Rio Grande Valley have long been recognized as distinct from populations elsewhere in Texas, which are assigned to the species *Siren intermedia* (Lesser Siren). At one time, these salamanders were regarded as a subspecies of *S. intermedia* and were designated *S. i. texana* (the Rio Grande Siren). Other researchers considered them to represent populations of the southeastern species *Siren lacertina* (Greater Siren). But there is a widespread consensus among Texas herpetologists that sirens in the Lower Rio Grande Valley, which are quite different from Greater Sirens, represent a distinct species. Unfortunately, under the rules of the ICZN, the last name applied to them is accepted as the most appropriate scientific name; therefore, the name *Siren lacertina* has been widely applied to salamanders in these populations, despite well-understood morphological differences between Rio Grande Sirens and southeastern Greater Sirens. Consequently, we chose to informally recognize the siren populations in the Lower Rio Grande Valley as *Siren* sp. "Rio Grande."

KEYS

A dichotomous key is an identification tool used in science. We use keys to separate genera and species, usually based on characteristics including morphology, color, and pattern. Constructing a usable key can be a daunting task. We hope that the ones presented here are user-friendly and that the reader will find them to be applicable in the laboratory and the field. They are dichotomous identification keys subdivided by major taxonomic groups, and the terms used in the keys were designed for the benefit of less practiced students of herpetology; the majority of terms are in standard use by herpetologists. The keys assume that the reader understands the difference between amphibians and the rest of the animal world, as well as between salamanders, and frogs and toads. Compare the keys to the drawings to better understand the names of the morphologies.

Our keys (adapted from Dixon 2000) are designed for determinations at the species level, but the reader may still find the necessary information lacking. There can be several reasons for this. Our key was developed for living animals or for freshly killed animals found DOR. Therefore, there may be problems using the keys to identify preserved animals, because preservatives make it difficult to move limbs, affect the size of the animal, change patterns and colors, and distort their natural stature. Toads are a particular problem for any amphibian key. Some toad species freely hybridize, and this makes using a key problematic. Hybridization can cause particular havoc with key characteristics such as the shape of the parotoid glands, so the reader should key the specimen as closely as possible, then go back to the literature to confirm the findings. To further complicate this problem, the first-generation (F1) offspring of these animals are often fertile.

KEY TO THE SALAMANDERS OF TEXAS

This key was adapted from Dixon (2000).

1. a. Hind limbs present. 3
   b. Hind limbs absent 2
2. a. Costal grooves, 34–36, between limbs and anus; olive gray above with scattered black spots; belly with numerous light spots (see Figure 2). . . . . Lesser Siren
   Siren intermedia
   b. Costal grooves, 36–38, between limbs and anus; gray or brownish gray above with tiny black spots; belly gray
   . . . . .Rio Grande Siren
   Siren species "Rio Grande"
3. a. Hind toes, 4 or 5; body not eellike. 4
   b. Hind toes, 3; body eellike . . . . . . Three-toed Amphiuma
   Amphiuma tridactylum
4. a. Hind toes, 4 5
   b. Hind toes, 5 6
5. a. Aquatic salamander with 3 pairs of external gills
   . . . . Gulf Coast Waterdog
   Necturus beyeri
   b. Very small terrestrial brown salamander without external gills and with 4 toes on each limb
   . . . . Dwarf Salamander
   Eurycea quadridigitata
6. a. Costal grooves absent or indistinct; top of head rough with numerous low ridges, including canthus rostralis (area in front of the eye to the nostril) 7

b. Costal grooves well developed; top of head smooth, without ridges 8

7. a. Black belly spots large, about the size of the eye; dorsum never reddish or with red spots
. . . . .Black-spotted Newt
Notophthalmus meridionalis

b. Black belly spots small, smaller than the eye; dorsum reddish, often with red spots . . . . Eastern Newt
Notophthalmus viridescens

8. a. Nasolabial groove (depression or trough extending from the nostril to the edge of the upper lip) present if external gills absent; if external gills present, then with 3 gill slits (Figure 2) 9

b. Nasolabial groove absent; if external gills present, then with 4 gill slits. 23

9. a. Belly uniformly light; tail laterally compressed 10

b. Belly solid black or black with light flecks; tail round in cross section . . . . Western Slimy Salamander
Plethodon albagula

10. a. External gills and 3 gill slits; no nasolabial groove. 11

b. External gills absent; nasolabial groove present
. . . . Southern Dusky Salamander
Desmognathus auriculatus

11. a. Costal grooves, 13 or more; interorbital distance (distance between the eyes) a maximum of 6 eye diameters
. . . . 12

b. Costal grooves, 12 or fewer; interorbital distance a minimum of 10 eye diameters. 15

12. a. Vestigial eyes; dark eye spots under the skin. 13

b. Functional eyes, but reduced in size
. . . . .Comal Blind Salamander
Eurycea tridentifera

13. a. Tail fin well developed, Hays and Comal Counties
. . . . 14

b. Tail fin not well developed, Barton Springs, Travis County
. . . . Austin Blind Salamander
Eurycea waterlooensis

14. a. Adpressed* limbs overlap 1 costal groove
. . . . Blanco Blind Salamander
Eurycea robusta

b. Adpressed limbs overlap 5 or more costal grooves
...........................Texas Blind Salamander
Eurycea rathbuni

15. a. Eye lacking dark lens ring.........................16
    b. Eye with dark lens ring 21

16. a. 4 or 5 costal grooves between adpressed limbs; 14–17 costal grooves 17
    b. 3 or fewer costal grooves between adpressed limbs; 13–14 costal grooves ....Valdina Farms Salamander
    Eurycea troglodytes

17. a. Eye diameter 20%–25% of the interorbital distance
   18
    b. Eye diameter half the interorbital distance 19

18. a. Salado Springs, Bell County ....Salado Salamander
    Eurycea chisholmensis
    b. Comal, Kerr, Hays, and Kendall Counties
    ....Cascade Caverns Salamander
    Eurycea latitans

19. a. Barton Springs, Travis County
    .... Barton Springs Salamander
    Eurycea sosorum
    b. Springs of Bexar, Kendal, and Blanco Counties 20

20. a. Helotes Springs, Leon Springs, and Mueller's Spring, Bexar and Kendall Counties.........Texas Salamander
    Eurycea neotenes
    b. Springs of northeastern Kendall and southern Blanco Counties.....................Fern Bank Salamander
    Eurycea pterophila

21. a. Found in springs north of Hays County.............22
    b. San Marcos Springs, Hays County
    .... San Marcos Salamander
    Eurycea nana

22. a. South, Middle, and North Forks of the San Gabriel River, Williamson County
    ....Georgetown Salamander
    Eurycea naufragia
    b. Springs of western Travis and Williamson Counties
    ....Jollyville Plateau Salamander
    Eurycea tonkawae

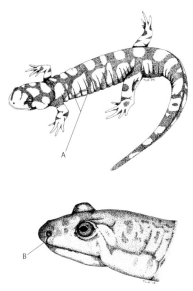

Figure 2. Basic salamander structures. *A*, costal grooves; *B*, nasolabial groove.

23. a. External gills present or absent; if external gills present, then gill rakers present 24
    b. External gills present, but without gill rakers
       . . . . Southern Dusky Salamander larvae
       Desmognathus auriculatus
24. a. Body with small light flecks (at least laterally) or lichen-like markings or totally dark 25
    b. Body with clearly defined spots, bars, or blotches 26
25. a. Costal grooves, 10; head broad and flat, considerably wider than neck; body short and stout; tail also short
       . . . . Mole Salamander
       Ambystoma talpoideum
    b. Costal grooves, 14; head not broad, little wider than neck; body moderately slender
       . . . . Small-mouthed Salamander
       Ambystoma texanum

26. a. Body with pale yellow to orange markings 27
    b. Body with pale metallic-white markings
       . . . . Marbled Salamander
       Ambystoma opacum
27. a. Yellowish to orange round spots arranged in 2 rows
       dorsally. . . . . Spotted Salamander
       Ambystoma maculatum
    b. Yellow blotches or bars extending onto sides and often
       onto belly 28
28. a. Pale spots or blotches on body, 15 to 58 (avg. 30)
       . . . . Eastern Tiger Salamander
       Ambystoma tigrinum
    b. Pale spots or blotches on body large, 6 to 36 (avg. 17)
       . . . . Barred Tiger Salamander
       Ambystoma mavortium

*Note: To configure adpressed limbs in animals with 4 limbs, the forelimbs are laid backward along the side of the body, and the hind limbs are laid forward along the side of the body, both at full length. It is difficult or impossible to get properly adpressed limbs on a preserved specimen without damaging it.

KEY TO THE FROGS OF TEXAS

This key was adapted from Dixon (2000).

1. a. Pupil of eye vertical 2
   b. Pupil of eye round or horizontal 6
2. a. Tongue attached in front of mouth 3
   b. Tongue attached in rear of mouth
      . . . . Mexican Burrowing Toad
      Rhinophrynus dorsalis
3. a. Pectoral glands (chest glands) absent 4
   b. Pectoral glands present . . . . Hurter's Spadefoot
      Scaphiopus hurterii
4. a. Knob-like process (boss) absent between the eyes 5
   b. Boss present between the eyes . . . . Plains Spadefoot
      Spea bombifrons
5. a. Pigmented heel pad on the hind foot rounded, about as
      wide as long . . . . Mexican Spadefoot
      Spea multiplicata

b. Pigmented heel pad on the hind foot elongated, about twice as long as wide . . . . .Couch's Spadefoot
Scaphiopus couchii

6. a. Tympanum (external ear drum) present (see Figure 3) 9

b. Tympanum absent 7

7. a. Heel pad, 1; light middorsal line absent; in open mouth, when looking down the throat, there are no smooth dermal ridges across the palate in front of the throat 8

b. Heel pads, 2; light middorsal line present; in open mouth, when looking down the throat, there are 2 dermal ridges across the palate in front of the throat . . . . . . Sheep Frog
Hypopachus variolosus

8. a. Belly immaculately white to cream . . . . Great Plains Narrow-mouthed Toad
Gastrophryne olivacea

b. Belly darkly mottled. . . . Eastern Narrow-mouthed Toad
Gastrophryne carolinensis

9. a. Horn-like projections, 2, on rear of tongue 10

b. No horn-like projections on rear of tongue 18

10. a. Dorsolateral folds (elongated folds on side of body) absent (see Figure 3) 11

b. Dorsolateral folds present 12

11. a. Distance from heel to knee about equal to the distance from heel to first toe; distance from corner of mouth to tip of snout equal to the width of the head at the posterior (back) edge of the tympanum (see Figure 3) . . . . .Pig Frog
Lithobates grylio

b. Distance from heel to knee about equal to the distance from heel to second toe; distance from corner of mouth to tip of snout about 24% greater than the width of the head at the anterior (forward) edge of the tympanum . . . . American Bullfrog
Lithobates catesbeianus

12. a. Dorsolateral fold extending entire length of body (may be broken posteriorly); dorsal pattern of spots 13

b. Dorsolateral fold extends ⅔ length of body; general color brown . . . . Green Frog
Lithobates clamitans

13. a. Spots between dorsolateral folds rounded; folds consistent in width throughout their length. 14
    b. Spots between the dorsolateral folds in 2 rectangular rows; folds are much broader anteriorly than posteriorly
    . . . . Pickerel Frog
    Lithobates palustris

14. a. Dorsal spots (spots on the back) not bordered by a circle of white. 15
    b. Dorsal spots bordered by a circle of white
    . . . . Crawfish Frog
    Lithobates areolatus

15. a. Dorsolateral folds continuous to the hind limb insertions (where the leg goes into the body) 16
    b. Dorsolateral folds interrupted at the rear. 17

16. a. Dark dorsal spots usually with light borders; rear thigh has a pattern of discrete dark spots; ear drum light spot absent . . . . Northern Leopard Frog
    Lithobates pipiens
    b. Dark dorsal spots without light borders; rear thigh pattern of dark reticulating lines; ear drum light spot present; dark snout spot usually absent; external vocal sacs*
    . . . . Southern Leopard Frog
    Lithobates sphenocephalus

17. a. Well-defined whitish upper-lip stripe from rear angle of jaw almost to tip of snout; white eardrum and dark snout usually present; pattern of indistinct, brownish, reticulating lines on posterior thigh . . . . . Plains Leopard Frog
    Lithobates blairi
    b. White stripe on upper lip, when present, not well defined behind eye; white eardrum and dark snout spots usually absent; pattern of heavy black reticulating lines on posterior thigh. . . . . Rio Grande Leopard Frog
    Lithobates berlandieri

18. a. Parotoid glands (large, swollen toxin gland on neck behind eye) and warty skin present (see Figure 3) 19
    b. Parotoid glands and warty skin not present 28

19. a. Parotoid glands triangular in shape, as long as head
   20
    b. Parotoid glands not triangular in shape; not as long as head. 21

20. a. Color yellowish-brown to brown; adults large, greater than 84 mm (3.3 in.) from snout to vent
. . . . .Cane Toad
Rhinella marina

   b. Color green to olive-green; adults less than 60 mm (2.4 in.) from snout to vent . . . . .Green Toad
Anaxyrus debilis

21. a. Parotoid glands circular or tear-shaped. 22

   b. Parotoid glands elongated; from nearly 2 to more than 3 times longer than wide 23

22. a. Parotoid glands tear-shaped; cranial crests prominent
. . . . Gulf Coast Toad
Incilius nebulifer

   b. Parotoid glands circular; cranial crests absent or poorly defined . . . . Red-spotted Toad
Anaxyrus punctatus

23. a. Supraorbital crests (crest above the eyes) parallel, not united into a boss (bump) at the level of the front margin of the eyes. 24

   b. Supraorbital crests united into a boss at the level of the front margin of the eyes . . . . Great Plains Toad
Anaxyrus cognatus

24. a. Cranial crests prominent (see Figure 3) 25

   b. Cranial crests absent or poorly defined . . . . . Texas Toad
Anaxyrus speciosus

25. a. Parotoid glands closest together at their midpoint
. . . . 26

   b. Parotoid glands closest together at their anterior. 27

26. a. Dark spots present on chest and, at times, on body
. . . . Fowler's Toad
Anaxyrus fowleri

   b. Either a single dark spot present on chest or dark spots absent; a few dark spots possibly present on throat of male
. . . . Woodhouse's Toad
Anaxyrus woodhousii

27. a. Femoral warts (thigh warts) small and tibial warts (shin warts) large; postorbital (behind the eye) and supra-orbital crests about equal in size . . . . American Toad
Anaxyrus americanus

b. Both femoral and tibial warts small and about equal in size; postorbital crests conspicuously larger than supraorbital crests .... Houston Toad
Anaxyrus houstonensis

28. a. Toes with intercalary cartilage (in hylids, the cartilage between the bones in the toes) 29

b. Toes without intercalary cartilage or bone 39

29. a. Toe pads greatly reduced in size, little wider than digits (toes) 30

b. Toe pads large, distinctly wider than digits.......... 33

30. a. Webs between toes poorly developed or reduced in size (see Figure 3) 31

b. Webs between toes well developed, extending nearly to toe tips .... Northern Cricket Frog
Acris crepitans

31. a. Dorsal pattern of spots or blotches 32

b. Dorsal pattern of 3 dark lines...... Upland Chorus Frog
Pseudacris feriarum

32. a. Distinct, uninterrupted dark line from eye to midbody on each side; broken lines or rows of spots dorsally .... Spotted Chorus Frog
Pseudacris clarkii

b. No uninterrupted dark line from eye to midbody; body possibly spotted, unicolored, or blotched .... Strecker's Chorus Frog
Pseudacris streckeri

33. a. Eardrum much smaller than eye................... 34

b. Eardrum about equal in size to eye... Mexican Treefrog
Smilisca baudinii

34. a. Black-bordered pale spot below eye................ 35

b. No black-bordered pale spot below eye............. 36

35. a. Slow trilling call, about 24 pulses per second .... Gray Treefrog
Hyla versicolor

b. Fast trilling call, about 50 pulses per second .... Cope's Gray Treefrog
Hyla chrysoscelis

36. a. Top of body uniform in color or with smaller, darker spots or markings.............................. 37

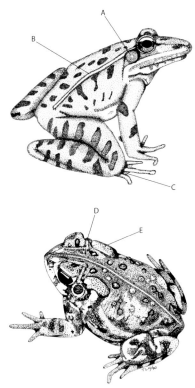

Figure 3. Basic structures of frogs and toads. *A*, tympanum (eardrum); *B*, dorsolateral fold; *C*, webbing between toes; *D*, cranial crest; *E*, parotoid gland.

      b. Top of body with an X-shaped pattern . . . .Spring Peeper
                                             Pseudacris crucifer
37. a. Top of body unicolor or with only a few dark spots. . . 38
      b. Top of body with many dark spots . . . .Canyon Treefrog
                                              Hyla arenicolor
38. a. Diffuse white stripe from forelimb insertions; adults usually less than 31 mm (1.2 in.) from snout to vent
          . . . . .Squirrel Treefrog
                                              Hyla squirella

b. Prominent white stripe from lips to midbody or slightly beyond, seldom reaching to hind limb insertions ; adults bright green; usually more than 36 mm (1.4 in.) from snout to vent .... Green Treefrog
Hyla cinerea

39. a. Tips of digits expanded, frequently T-shaped 40
    b. Tips of digits not expanded
    .... Mexican White-lipped Frog
    Leptodactylus fragilis

40. a. Vomerine teeth (teeth located on paired bones on roof of mouth) absent 41
    b. Vomerine teeth present .... Barking Frog
    Craugastor augusti

41. a. Dorsal markings of numerous irregular lines, spots, or blotches; length of forelimb usually greater than the foot with tarsus (flat of foot) 42
    b. Dorsal markings of a few poorly defined spots; length of forelimb usually less than the foot with tarsus
    .... Rio Grande Chirping Frog
    Syrrhophus cystignathoides

42. a. Dark bar present between eyes... Spotted Chirping Frog
    Syrrhophus guttilatus
    b. Dark bar absent between eyes ......Cliff Chirping Frog
    Syrrhophus marnockii

*Note: Vocal sacs usually cannot be seen in preserved specimens.

GENERIC KEY TO AMPHIBIAN LARVAE

This key was adapted from Dixon (2000) and prepared by John H. Malone.

1. a. External, feathery gills present....................... 2
   b. No gills, anterior body rotund with tail.............. 9
2. a. Hind limbs absent ....Siren
   b. Hind limbs present............................. 3
3. a. 3 toes.................................. *Amphiuma*
   b. 4–5 toes on hind limbs............................ 4

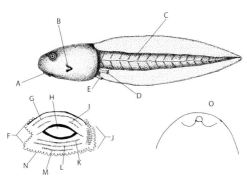

Figure 4. Tadpole basics. *A*, mouth; *B*, spiracle; *C*, tail musculature; *D*, limb bud; *E*, anus; *F*, emarginate papillae; *G*, first anterior tooth row; *H*, mouth; *I*, second anterior tooth row; *J*, non-emarginate papillae; *K*, first posterior tooth row; *L*, second posterior tooth row; *M*, third posterior tooth row; *N*, papillae; *O*, oral flap.

4. a. 4 toes on hind limbs 5
   b. 5 toes on hind limbs 6
5. a. Dorsal fin ending approximately perpendicular to vent
       . . . . *Necturus*
   b. Dorsal fin extending well past vent onto body
       . . . . *Eurycea*
6. a. 3 gill slits 7
   b. 4 gill slits 8
7. a. Soles of feet keratinized . . . . *Desmognathus*
   b. Soles of feet not keratinized . . . . *Notophthalmus*
8. a. Dorsal fin ending perpendicular to vent . . . . *Eurycea*
   b. Dorsal fin extending well past vent . . . . *Ambystoma*
9. a. Hard, horny mouth absent, no anterior or posterior teeth
       rows 10
   b. Hard, horny mouth with anterior and posterior teeth
       rows 12
10. a. Barbel projections on anterior region of body; 2 spiracles
        positioned midlaterally . . . . *Rhinophrynus*
    b. No barbel projections on anterior body 11
11. a. Oral flap with medial gap; marginal edges of mouth flaps
        with papillae bumps . . . . *Hypopachus*
    b. Oral flap with little or no medial gap; marginal edges of
        mouth flaps smooth . . . . *Gastrophryne*

12. a. Eyes dorsal..13
    b. Eyes lateral..17
13. a. Marginal papillae emarginated 14
    b. Marginal papillae not emarginated.................15
14. a. Distinct posterior papillary gap
        .... *Anaxyrus, Incilius,* or *Rhinella (Bufo)*
    b. No posterior papillary gap .... *Lithobates (Rana)*
15. a. Distinct anterior papillary gap..................... 16
    b. Little or no anterior papillary gap 18
16. a. Found in Cameron, Hidalgo, and Starr Counties
        ....*Leptodactylus*
    b. Found in West Texas ....*Hyla*
17. a. Body wider posteriorly than anteriorly; up to 35 mm
        (1.4 in.) total length...................... *Scaphiopus*
    b. Body wider anteriorly than posteriorly; up to 100 mm
        (3.9 in.) total length ....*Spea*
18. a. 2 posterior teeth rows............................. 19
    b. 3 or more posterior teeth rows 20
19. a. Tail with black tip........................... *Acris*
    b. Tail without black tip......................*Pseudacris*
20. a. Upper jaw with long processes that project laterally;
        found in Cameron and Hidalgo Counties
        ....*Smilisca*
    b. Upper jaw with short processes that project downward;
        widespread throughout the state
        ....*Hyla* or *Pseudacris*

SYSTEMATIC ACCOUNTS

The following systematic accounts describe 2 orders, 15 families, and 72 species. Orders and families are described briefly and in general terms. In the species accounts, there are detailed sections for size, description, voice (if applicable), similar species, distribution (with maps), natural history, reproduction, subspecies (if applicable), and comments and conservation.

Note: Parentheses around the name of a describing author indicate that the genus name has changed since the original description was published. If an author who named a species placed it in the same genus used today, the author's name is given without parentheses.

ORDER CAUDATA: SALAMANDERS

All Texas salamanders tend to have "lizard-like" bodies with legs that are usually longer than the body is wide; all have eyelids; all except newts (family Salamandridae) are distinguished by costal grooves; the vent is longitudinal; and all have tails as adults. Most have 5 digits on the hind feet (when present), but there are a few genera with 4 toes on the hind feet, and one with only 3 digits. Most have only 4 toes on the front feet. Depending on the species, amphiumas have 1, 2, or 3 toes on the front feet. Except in paedomorphic or neotenic individuals, transformed adults do not have external gills. Like frogs, salamanders, with a few exceptions, usually lay eggs in water and hatch gilled larvae that transform into a nongilled adult stage.

FAMILY AMBYSTOMATIDAE: MOLE SALAMANDERS

The family Ambystomatidae is a New World family found from southeastern Alaska, and in Canada from southern Labrador, to the southern part of the Mexican Plateau. Two genera, *Ambystoma* and *Dicamptodon*, are recognized, with about 30 living species. This family has the following characteristics: the absence of nasolabial grooves, transformed adults having lungs, and a total length at maturity of approximately 200 mm (7.9 in.), but sometimes reaching in excess of 340 mm (13.4 in.). Larvae have broad heads, 3 pairs of bushy gills with well-developed rachises, and broad caudal fins that extend well onto the back. The prevomerine teeth occur in a transverse row that crosses the palate near the posterior margins of the internal nares (nostril openings).

Spotted Salamander
Ambystoma maculatum,
(Shaw, 1802)

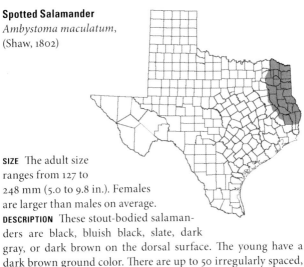

SIZE The adult size ranges from 127 to 248 mm (5.0 to 9.8 in.). Females are larger than males on average.

DESCRIPTION These stout-bodied salamanders are black, bluish black, slate, dark gray, or dark brown on the dorsal surface. The young have a dark brown ground color. There are up to 50 irregularly spaced, round, yellow or orange spots, beginning just behind the eyes and extending to the tip of the tail. The spots are usually in 2 rows along the dorsolateral surface of the body. In the southern part of their range, the spots on the head and neck are orange. The ventral surface of the animal is usually slate gray. Some individuals may be unspotted. These salamanders have 12 costal grooves. Spotted Salamanders have well-developed lungs, but

Spotted Salamander, San Augustine County.

pulmonary respiration contributes only 30 percent of the total gas exchange; cutaneous respiration contributes the balance.

VOICE Vocalization has been characterized as *tic-tic-tic*.

SIMILAR SPECIES In Texas, the only 2 terrestrial salamanders that grow larger than the Spotted Salamander are the Eastern Tiger Salamander, *Ambystoma tigrinum*, and the Barred Tiger Salamander, *Ambystoma mavortium*. The Barred Tiger Salamander and the Spotted Salamander do not occur in sympatry (the same area). The Eastern Tiger Salamander can readily be distinguished by its pattern of irregularly scattered spots and its larger, broader head.

DISTRIBUTION Spotted Salamanders are widely distributed across the eastern half of North America, from south-central Ontario to Nova Scotia, south to Georgia and west to eastern Texas. They are absent from prairies and open areas within their range. In Texas, the Spotted Salamander is found from the northeast corner of the state (Red River and Bowie Counties) southward to Jasper and Newton Counties, predominately east of the Neches River drainage. It is apparently absent from the counties of extreme southeast Texas, where forests give way to coastal prairies.

NATURAL HISTORY These salamanders are typically found on hillsides in moist hardwood forests and are most frequently encountered beneath or within decaying logs. They have also been recorded under stones, boards, and leaf litter. While most of their lives are spent underground in burrows, they can be found on

Spotted Salamander, San Augustine County.

Spotted Salamander, Red River County.

the surface during wet weather, particularly during the breeding season, when they migrate to breeding ponds. The home ranges of these salamanders vary between 3.3 and 29.4 square meters (29.4 and 318.5 square feet), with a mean of 9.8 square meters (106.6 square feet). Live Spotted Salamanders have been found at temperatures down to 0°C (32°F). In the northern part of their range, they are most active after the first spring thaw, as early as January and February, while in Texas, they have been found surface active from December through February. In Texas, they may be encountered under cover objects as late as the end of April or early May, as long as moisture and cool temperatures continue. But once the Texas sun dries the landscape, these salamanders retreat to their subterranean burrows. Juveniles and adults eat a variety of invertebrates, such as mollusks, earthworms, centipedes, millipedes, spiders, and insects. They may feed from their underground retreats or forage during wet weather. Their eggs and larvae may be preyed on by tadpoles, small fish, various aquatic invertebrates, sandpipers, and other amphibian larvae (especially Marbled Salamander larvae). Predators of adults include raccoons, opossums, weasels, and minks.

REPRODUCTION Spotted Salamanders breed in the very early spring in the northern part of their range, usually during January and February. In the southern part of their range, these salamanders breed during the first warm rains that occur from December through February. Heavy rain and warming temperatures initiate the migration of these salamanders to breeding ponds. The eggs are deposited in a single mass suspended on vegetation. The duration of development of the eggs is 43 days, and the larvae take 87 days to develop. Larvae rarely overwinter in breeding ponds.

COMMENTS AND CONSERVATION The Spotted Salamander is on the TPWD's Black List. Spotted Salamanders seem particularly tied to mature hardwood forests, and their populations in Texas are therefore susceptible to local extirpation following the harvesting of these forests. Inundation of forests after reservoir construction has also led to the loss of many populations.

Barred Tiger Salamander
Ambystoma mavortium,
Baird, 1850

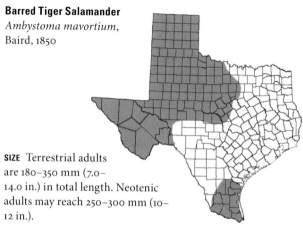

SIZE Terrestrial adults
are 180–350 mm (7.0–
14.0 in.) in total length. Neotenic
adults may reach 250–300 mm (10–
12 in.).

DESCRIPTION Terrestrial adults have a black
or gray-black dorsal ground color that is covered with broad,
vertical pale yellow, yellow, or olive bars or large spots that extend from the lower sides and belly to near the middorsal region, with some rounded or irregular pale yellow, yellow, or
olive markings interspersed between the vertical bands. In recently transformed individuals or neotenic adults, the ground
color may be suffused with pale green to olive, with black spots
covering the dorsal surface. Terrestrial adults are stout bodied
with small protruding eyes and broad, flattened heads that ter-
minate in a rounded snout with a large gaping mouth. They also
have laterally compressed tails that are rounded below, sharp-
edged above, and flattened near the tip; the tail is usually as long
as the head and the body unless it has been broken. Neotenic
adults, or axolotls, usually retain many larval characteristics,
such as a broad flat head, stout body, and broad fins. The dorsal
fin extends from the back of the head, becoming wider behind
the vent, and the ventral fin extends to the anus. Both forms
have long sturdy legs with 4 toes on the front feet and 5 long,
webbed, flattened toes on the back feet. Tubercles used for digging are present on the underside of the front and rear feet, with
a horn-like tip on each toe. There is a well-developed gular fold.
When the limbs are adpressed, the toes overlap 2–4 intercostal
folds. There are 11–14 costal grooves. Males have distinctively
swollen cloacae when breeding.

SIMILAR SPECIES This is a highly variable species, but markings are generally larger and fewer than the light markings on the Eastern Tiger Salamander (*Ambystoma tigrinum*). Neotenic adults may have large external gills or gill buds. The Eastern Tiger Salamander does not have a neotenic form in Texas.

DISTRIBUTION This species of salamander has the largest known range in North America, ranging from western Canada southward through Montana, Wyoming, Utah, and Arizona into the western mainland of Mexico. From western Mexico, it then is found eastward into central Mexico. East of the Rocky Mountains, it ranges from the southern tip of Texas northward into Oklahoma, Kansas, Nebraska, the Dakotas, and southern Canada. Disjunct populations have been found west of the Rocky Mountains. *Ambystoma mavortium* has been found vertically from sea level to about 3,350 m (10,991 ft.) in elevation.

NATURAL HISTORY There have been 4 adult morphs recognized in this species. They include typical metamorphosed adults, cannibalistic metamorphosed adults, typical gilled adults, and cannibalistic gilled adults. In West Texas, 2 types of typical adults occur. They differ in color pattern and clutch size. One morph inhabits seasonally temporary ponds, transforming at a relatively small size when immature, while the other morph in-

Barred Tiger Salamander, Kent County.

habits highly eutrophic playas, typically transforming when sexually mature. Terrestrial or fossorial adults (those adapted to digging) are nocturnal and may be found in a wide variety of habitats. They are generally found close to quiet water habitats used for breeding, which include riparian deciduous forests, ponds, reservoirs, lakes, temporary pools, arid sagebrush plains, rolling grasslands, mountain meadows, coniferous forests and woodlands, open fields and brushy areas, alpine and subalpine meadows, grasslands, semideserts, and deserts. They prefer areas with sandy or otherwise friable soil so that they can easily burrow near breeding sites. Terrestrial adults have been found under objects and in crawfish or mammal burrows near water. They become active on the ground surface during and after periods of heavy summer and fall rains. The primary diet of the larvae is zooplankton, but they may also feed on nematodes, insects, snails, clams, cladocerans, copepods, ostracods, fairy shrimp, crayfish, and amphibian eggs and larvae. Cannibalistic larvae feed on larger prey, such as other salamander larvae and frog tadpoles, and often consume larvae nearly as large as themselves. Adults eat a variety of organisms, including beetles, grasshoppers, caterpillars, grubs, worms, snails, and tadpoles. Egg predators may include other Tiger Salamander larvae, dragonfly nymphs, caddis flies, diving beetles, and garter snakes. Many bird species, such as bitterns, Killdeers, grackles, owls, jays, and herons, prey on larvae and adults. Bobcats, coyotes, badgers, and snakes have been observed feeding on adults also. Automobile traffic can kill large numbers of adults and juveniles on rainy nights as the salamanders move about in search of food or breeding sites. Antipredator responses include defensive posturing with the head down and the tail waving up in the air and excreting milky noxious substances. These are long-lived salamanders and may reach 25 years in captivity. The authors have maintained terrestrial adults for more than 16 years in captivity.

REPRODUCTION Breeding takes place from January to May in temporary and permanent cattle tanks, subalpine lakes, ditches, and, rarely, streams. Barred Tiger Salamanders in the western United States breed most often in permanent water habitats, but rarely in sites with predatory fish. Females in some populations may be capable of long-term sperm storage. Eggs are laid singly, in very small clusters, or in linear strings on twigs, vegetation,

Barred Tiger Salamander, Stonewall County.

or detritus. Incubation period may be 1–2 weeks. Larvae are 75–125 mm (3–5 in.) at metamorphosis.

SUBSPECIES There are 5 subspecies of western Tiger Salamanders in this species complex. They include the Barred Tiger Salamander (*Ambystoma mavortium mavortium*), Blotched Tiger Salamander (*A. m. melanostictum*), Gray Tiger Salamander (*A. m. diabolia*), Arizona Tiger Salamander (*A. m. nebulosum*), and Sonoran Tiger Salamander (*A. m. stebbinsi*). These are part of a species complex that also includes the Eastern Tiger Salamander (*A. tigrinum*) and the California Tiger Salamander (*A. californiense).* To make matters worse, local populations may contain various morphs, including cannibalistic forms, neotenic forms, or transformed adults with sexually immature larvae that may occur as either typical or cannibalistic morphs.

COMMENTS AND CONSERVATION The TPWD still considers the Barred Tiger Salamander a subspecies to the Tiger Salamander, and has therefore placed it on the White List. Larvae are used as fish bait, and commercial bait collectors have introduced nonnative subspecies into many regions of the western United States; Texas is no exception. These introduced specimens have disrupted efforts to determine the original distribution of species and have made identification more challenging. Loss of wetland habitats and the introduction of fish have had a negative effect on some populations.

Marbled Salamander
Ambystoma opacum,
(Gravenhorst, 1807)

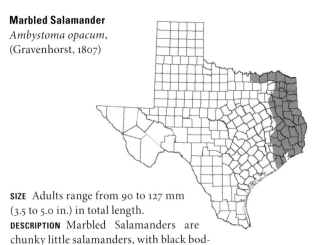

SIZE Adults range from 90 to 127 mm
(3.5 to 5.0 in.) in total length.

DESCRIPTION Marbled Salamanders are
chunky little salamanders, with black bodies boldly marked with white (males) or grayish-white (females) crossbars. The light crossbars are variable: sometimes incomplete, running together, or enclosing dark spots. Occasionally, the crossbars will fuse laterally, forming stripes, with the occasional individual lacking crossbars and having a pattern consisting of only the 2 stripes. The belly is unmarked, plain black. Newly transformed individuals are quite different in appearance from older adults, having a pattern of scattered light flecks on a dull brown to black background, with the spots and flecks coalescing into the cross-banded adult pattern over a period of a few months. There are 11–12 costal grooves in this species. Marbled Salamanders possess well-developed lungs along with well-developed skin capillaries. This skin capillary network is somewhat more developed anteriorly than posteriorly, and the posterior ventral surfaces are the least vascularized. Respiration occurs through both lungs and skin.

SIMILAR SPECIES The Marbled Salamander is not easily confused with other salamanders in Texas.

DISTRIBUTION The Marbled Salamander is distributed throughout the forested regions of the eastern United States, from southern New England, Lake Michigan, and Lake Erie to northern Florida and west to southern Illinois, southeastern Oklahoma, and eastern Texas. In Texas, this salamander occurs chiefly east of the Trinity River. Along the Red River in North Texas, it has been found as far west as Grayson County, while in the southeastern portion of the state, it occurs as far west as Harris and Galveston Counties.

NATURAL HISTORY These salamanders occur in moist hardwood forests, ranging from riparian floodplain forests to (outside of Texas) the forested ridges of the Ozark Plateau and Appalachian highlands; they are absent from predominantly grassland habitats. They can tolerate drier hillsides than those inhabited by the Spotted Salamander. While they are surface active in wet, cool weather, they are most frequently encountered under surface objects such as logs, rocks, and boards. Surface activity typically corresponds to the fall breeding season. Hatchlings and small larvae tend to congregate in leaf litter in warm, shallow water during the day, but disperse more evenly at night throughout the ponds and feed in the water column. Adults feed on small worms, insects, slugs, snails, millipedes, centipedes, and spiders. The emerging hatchlings feed almost immediately on zooplankton. Younger larvae feed on the bottom of ponds; larger and older larvae actively forage at all levels of the water column. While they presumably feed primarily on aquatic invertebrates, they have been recorded as feeding upon the aquatic eggs of frogs and other species of salamanders. Although larvae are known to be aggressive and cannibalistic, they have been shown to be able to recognize kin: they are less aggressive and more

Marbled Salamander, San Augustine County.

Marbled Salamander, Tyler County.

submissive to siblings than to nonsiblings. Egg predation can be by beetles, salamanders, and frogs. Larvae are preyed upon by fishes, other amphibians, snakes, and wading birds as well as insects and invertebrates. Adults may be preyed upon by owls, raccoons, skunks, weasels, shrews, snakes, and other woodland predators. Both males and females exhibit defensive posturing by tail lashing, body coiling, and head-butting behaviors as well as by becoming immobile.

REPRODUCTION Reproduction occurs from September to October north of Texas and from October to December in the state, with annual variations tied to rainfall. After fertilization, the female deposits the eggs in a sheltered depression that will later fill with water. The female guards her clutch of 50–200 eggs until they are washed into aquatic sites by flooding. The female will occasionally agitate the eggs and move them about, presumably to prevent fungal growth. Egg guarding continues for 41–52 days. The larvae are positively phototactic (move toward light) until their hind limbs are fully developed, when they become negatively phototactic (moving away from light). They can frequently be observed swimming in pools in flooded forests at night.

COMMENTS AND CONSERVATION The Marbled Salamander is on the TPWD's Black List. Their populations are susceptible to local extirpation when forests are harvested. Inundation of riparian forests to produce large reservoirs has also undoubtedly destroyed many populations.

Mole Salamander
Ambystoma talpoideum,
(Holbrook, 1838)

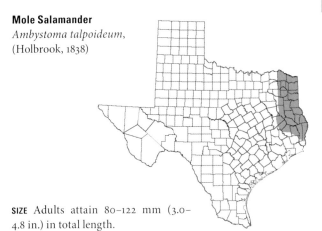

SIZE Adults attain 80–122 mm (3.0–4.8 in.) in total length.

DESCRIPTION *Ambystoma talpoideum* may exist as either a sexually active neotenic adult or a terrestrial adult. The dorsal ground color of the terrestrial form varies from light brown through light bluish-gray and dark gray to almost black. There may be conspicuous or inconspicuous small bluish-white to light gray specks along the lateral sides or on the dorsal surface. Along the upper crest of the tail, a pale area may be present. The ventral surface is gray with light blotches. The coloration of the neotenic adult is similar to that of the terrestrial adult. The terrestrial form has a large, prominent, rounded head; a short, stocky body; a relatively short tail; and rather short, large legs. The neotenic adult has a laterally flattened tail and a flattened head with external gills that may be almost nonexistent or prominent. There are 10 or 11 costal grooves. Sexual dimorphism is evident: females have smooth tails, whereas males have rugose tails. Males of both forms have swollen cloacae.

SIMILAR SPECIES The Small-mouthed Salamander (*Ambystoma texanum*) has a smaller head in relation to its body and a much longer tail.

DISTRIBUTION The Mole Salamander occurs along the southern Atlantic Coastal Plain and throughout much of the Gulf Coastal Plain of the southeastern United States, from South Carolina southward into northern Florida and westward into extreme East Texas; the range also extends northward along the Mississippi River Valley into southern Illinois. There are scattered dis-

Mole Salamander, Houston County.

junct populations west and north of the Atlantic Coastal Plain and the Gulf Coastal Plain.

NATURAL HISTORY The habitat of adults of the Coastal Plain areas are lowlands and valleys with extensive floodplain forests, in and near tupelo gum and cypress ponds. Populations outside the Coastal Plain are found in upland hardwood forests or mixed pine-hardwood forests with fishless permanent or semipermanent pools. Populations may consist of terrestrial adults or a mixture of neotenic and terrestrial adults. Terrestrial adults prefer areas where ponds are seasonal or semipermanent, while neotenic adults are more common near permanent ponds, including gravel pits and roadside ditches. In Texas, *Ambystoma talpoideum* has been found in the floodplains of lowland forested areas, but is most abundant in upland pine-hardwood forested areas. Terrestrial adults seek out moist environments with soft earth for burrowing, often taking advantage of animal burrows. Most of their time is spent underground, and adults usually emerge only for breeding. Terrestrial forms are occasionally encountered under debris such as logs or other objects in damp places, while both forms are found more frequently at their breeding sites. During daylight, neotenic and larval forms remain hidden in vegetation, bottom-resting debris, and leaf litter in their aquatic environment. The larvae's diet includes pond invertebrates and zooplankton such as fly larvae, chironomid larvae, ostracods, and cladocerans. Larvae may cannibalize eggs and consume other *Ambystoma* larvae. Adults feed on zooplankton, aquatic insects, tadpoles, and other salamander larvae. Predators of the eggs and larvae of *A. talpoideum* are primarily fish such as Bluegill and Green Sunfish. Adult predators include other ambystomatids, fish, wading birds, snakes, and small mammals. Antipredator responses include head butting, head-down posturing, tail lashing, biting, body flipping, and immobility. Noxious secretions that help protect them from predators are produced from well-developed parotoid glands and from glands along the top ridge of the tail. Individuals may live as long as 6–9 years in the wild.

REPRODUCTION Breeding, often in mass, occurs after or during intense periods of heavy, sustained rains. Cold temperatures do not appear to have an effect on breeding. Breeding in Texas usually occurs from December through February, whereas migrating salamanders may be observed as early as November. Neotenic adults breed earlier than terrestrial adults. These salamanders lay their eggs in small clusters floating in the water. There are 4–20 eggs per cluster. It may take as long a year for the larvae to transform into adults.

COMMENTS AND CONSERVATION These salamanders are on the TPWD's Black List. Mole Salamanders are locally common in Texas. Habitat destruction from the conversion of wetland-supporting upland forests into agricultural and urban areas has affected local populations. Clear-cutting of a mixed pine-hardwood forest adjoining a breeding pond has decreased a population of Mole Salamanders in Louisiana.

Mole Salamander, Houston County.

Small-mouthed Salamander
Ambystoma texanum,
(Matthes, 1855)

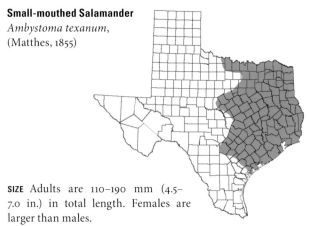

SIZE Adults are 110–190 mm (4.5–7.0 in.) in total length. Females are larger than males.

DESCRIPTION Adults may be brownish gray, grayish black, dark brown, or black, with light gray specks or lichen-like markings on the dorsal and lateral sides. These specks and lichen-like markings are much more prominent on the lower sides of the body. The ventral surface is similar in coloration to the dorsal surface, but the gray specks are less dense. The Small-mouthed Salamander has a large body and a conspicuously small head and mouth, with a tail as long as the head and body. The legs are relatively large, with 4 toes on the front feet, and 5 toes on the back feet. There are 14–15 costal grooves. Males have swollen papillose cloacae during the breeding season.

SIMILAR SPECIES The Mole Salamander may be confused with the Small-mouthed Salamander. The Mole Salamander has a much larger head in relationship to the rest of the body, and a tail shorter than the head and body. The Western Slimy Salamander (*Plethodon albagula*) has a nasolabial groove between its nostrils to the lip.

DISTRIBUTION Small-mouthed Salamanders range from northern Ohio westward into southeastern Nebraska and eastern Kansas, and southward into eastern Oklahoma and Texas, then eastward to eastern Kentucky, Tennessee, and western Alabama.

NATURAL HISTORY The Small-mouthed Salamander's habitat is bottomland forests associated with wetlands or adjoining floodplains. They have also been found in upland ponds, tall grassland prairies, and oak-hickory forests, especially in the western portion of their range. During the day, especially at breeding times, they spend their time in mammal burrows or crayfish burrows, and under logs, large rocks, and other cover objects near their breeding areas, which include fishless, semipermanent wetlands such as forested wetlands, flooded fields, prairie potholes, oxbows, roadside ditches, borrow pits, and stream pools. Excavation crews and farmers often plow them up. They emerge during rainy nights to search for breeding sites or to forage for food. Prey items include earthworms, isopods, centipedes, spiders, lepidopteran larvae, beetles and their larvae, and other insects. Larvae feed on zooplankton, isopods, amphipods, gastropods, copepods, and beetles while within the water column or from the surface of the leaf litter on the bottom. Small-mouthed Salamander predators include aquatic insects, Tiger Salamander larvae, garter snakes, and water snakes. Defense mechanisms include head posturing and tail waving. They lower their heads, curl their bodies, and stick their tails up in the air and wave them. The tail secretes a noxious substance that wards off many predators.

REPRODUCTION Breeding generally occurs immediately following winter or spring rains during mid-January to early March. Breeding migrations tend to be at night in rainy weather. Adults

Small-mouthed Salamander, Red River County.

Small-mouthed Salamander, Brazoria County.

engage in nonamplexing mass courtship. The eggs are deposited either singularly, in loose clusters, or in small masses on twigs, leaf petioles, leaves, grasses, other vegetation, and detritus in fishless ponds. Depending on water temperature, incubation lasts 2–8 weeks.

COMMENTS AND CONSERVATION This species is on the TPWD's Black List. Areas of urban development and clearing for agricultural purposes have reduced breeding sites; habitat destruction has decreased population numbers of Small-mouthed Salamanders in some areas. The authors found these salamanders to be quite common in the Houston, Dallas, and Fort Worth metro areas in the mid-twentieth century, but their populations seem to have declined dramatically with urbanization. Clear-cutting does not seem to affect Small-mouthed Salamander populations in southeast Texas.

Eastern Tiger Salamander
Ambystoma tigrinum,
(Green, 1825)

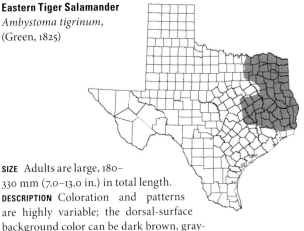

SIZE Adults are large, 180–330 mm (7.0–13.0 in.) in total length.

DESCRIPTION Coloration and patterns are highly variable; the dorsal-surface background color can be dark brown, grayish black, or black, and the dorsal surface is covered with a pattern of numerous large irregular light yellow, brownish-yellow, or yellowish-green blotches or spots. This pattern of blotches or spots extends down onto the sides. The ventral surface is marked with irregular yellow blotches on a darker background. The stout-bodied adult has a broad head with a rounded snout, a wide mouth, and small protruding eyes. Limbs and feet are short and stocky, with 4 toes on the front feet and 5 toes on the back feet. Digging tubercles are found on the underside of the front and rear feet, with one on each side of the rear undersurface. Rounded tails are usually as long as the head and body, and are flattened toward the tip in nonbreeding males. During breeding season, the male's tail becomes flattened from side to side. Breeding males have swollen cloacae.

SIMILAR SPECIES In Texas, only the Spotted Salamander (*Ambystoma maculatum*) can be confused with the Eastern Tiger Salamander. In *A. maculatum*, the light spots form 2 irregular rows down the dorsal surface, and the ventral surface is light gray to gray.

DISTRIBUTION This species occurs along the Atlantic and Gulf Coastal Plains from Long Island, New York, to eastern Texas. It is also found in Minnesota, Wisconsin, and Michigan, and then south into southeastern Oklahoma and northern Arkansas, and on into Mississippi, Alabama, Tennessee, and Kentucky. It is absent from mountainous regions. Disjunct populations are found scattered throughout the eastern third of Texas.

NATURAL HISTORY There are terrestrial adults and neotenic adults. Terrestrial adults are nocturnal and found under objects or in burrows near or in the following quiet-water habitats: bottomland deciduous forests, ponds, reservoirs, lakes, temporary pools, rolling grasslands, savannas, open fields, and brushy areas. Sandy and other friable soils are preferred for easy burrowing; they will also use mammal burrows near abundant breeding sites. Both the neotenic adults and terrestrial adults require fishless bodies of water in which to survive and to breed. Migrations generally occur during or shortly after rains. Neotenic adults are rare; the only known population occurs in Michigan. The diet of larvae and small neotenic adults includes aquatic insects, nematodes, terrestrial insects, snails, clams, cladocerans, copepods, earthworms, ostracods, leeches, amphipods, water mites, rotifers, fairy shrimp, crayfish, amphibian eggs and larvae, and the eggs and larvae of conspecifics. Food items of adults include insects, worms, snails, hatchling lizards, and young field mice. *Ambystoma tigrinum* eggs are preyed on by caddis flies, newts, and salamander larvae. Larvae predators include dragonfly naiads, caddis fly larvae, diving beetles, other salamander larvae, and garter snakes. Researchers have also reported that Eastern Tiger Salamander larvae are capable of matching their color to existing backgrounds, an ability that may reduce predation

Eastern Tiger Salamander, Robertson County.

Eastern Tiger Salamander, Lee County.

from birds and other sight-oriented predators. Large larvae and small adults may be preyed upon by Killdeers, night herons, bitterns, grackles, and jays. Bobcats and coyotes have been known to eat larvae trapped in drying ponds. Adults are taken by owls, snakes, and badgers. Bacteria also appear to be a problem for Eastern Tiger Salamanders. In certain populations, bacteria infestations may lead to major die-offs. Many adults are killed on roadways by automobiles during nightly excursions and migrations. Defensive posturing and milky excretions from the tail may be used to defend against predators.

REPRODUCTION Eastern Tiger Salamanders breed from November through May after heavy winter or spring rains. After a short courting process, the male deposits a spermatophore in front of the female, which she picks up with her cloacal lips. *Ambystoma tigrinum* deposits 400 eggs on average, in masses of 50 eggs, on twigs, weed stems, and other support structures in fishless ponds. Transformation occurs at a smaller size than in other species of tiger salamanders.

COMMENTS AND CONSERVATION The Eastern Tiger Salamander is on the TPWD's White List and is still considered a single species with the Barred Tiger Salamander. Populations of Eastern Tiger Salamanders in the southeastern United States have declined in many areas because of deforestation, loss of wetland habitats, and the introduction of fish into their breeding areas. Fish and wildlife managers should consider other options when deciding whether to stock naturally fish-free habitats.

FAMILY AMPHIUMIDAE: AMPHIUMAS

The family Amphiumidae is the smallest family of salamanders, containing a single genus with 3 species of *Amphiuma*, which are mainly restricted to the Gulf Coastal Plain and Mississippi Valley of the southeastern United States. Only one of these, *Amphiuma tridactylum*, is found in Texas. In Texas, they can be found in far eastern Texas from the Oklahoma border southward to the coast.

Adult amphiumas are long, dark-colored, eellike salamanders that may reach 1,060 mm (41.8 in.) in total length. They have reduced pelvic and pectoral girdles, and vestigial limbs. Their legs and arms are tiny and relatively useless. Adults lose their gills during transformation but have a single pair of gill slits. They also lack eyelids, possess costal grooves and lungs, but lack ypsiloid cartilage. Lungs are well developed. Amphiumas are aquatic and inhabit swamps, marshes, and other similar habitats.

Three-toed Amphiuma
Amphiuma tridactylum,
Cuvier, 1827

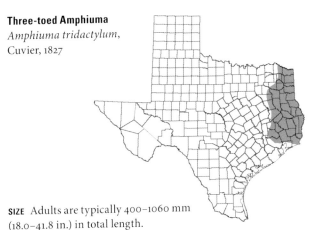

SIZE Adults are typically 400–1060 mm
(18.0–41.8 in.) in total length.

DESCRIPTION The dorsal surface is slate gray,
brown, or black, while the ventral surface is light gray. Where
the dorsal and ventral colors meet, a definite contrasting line
of demarcation occurs. The throat has a dark area on its center.
Amphiumas are large, eellike, elongated aquatic salamanders,
with relatively short tails and 4 tiny limbs, each having 3 minute
toes. A laterally compressed tail makes up 20 to 25 percent of the
total length. A single gill slit is present on each side and behind
the head, with no external gills present. These salamanders have
a rounded snout and small lidless eyes. Texas specimens tend to
have 62 costal grooves. Females have dark, smooth cloacal walls,
whereas males have light gray walls with papillose oval patches
and develop swollen cloacae during the breeding season.

VOICE Amphiumas have been reported to produce whistling
sounds.

SIMILAR SPECIES They are often confused with freshwater eels,
which have fins but not legs. Sirens (*Siren intermedia*) have external gills and only 2 front legs and no hind legs.

DISTRIBUTION The Three-toed Amphiuma is found throughout
the Mississippi Valley and the Gulf Coastal Plain. The range extends from southeastern Missouri, southeast into extreme western Kentucky and Tennessee, southward into eastern Arkansas,
most of Alabama, southwestern Georgia, and then west through
Louisiana into extreme southeastern Oklahoma and East Texas.

NATURAL HISTORY This strictly aquatic, nocturnal salamander inhabits permanent to semipermanent water habitats with abundant vegetation and bottom debris; examples include drainage ditches, marshes, bayous, swamps, sluggish streams, oxbows, irrigation ditches, and ponds. Amphiumas stay in their own burrows or crayfish burrows during the day and emerge at night to forage, with peak activity occurring 3–4 hours after sunset. Most activity occurs after heavy rains when temperatures exceed 5°C (41°F). They will hibernate in extremely cold weather. On nights with heavy rains, these salamanders may move overland, even crossing roads far from the water's edge. During extended dry spells, amphiumas can survive for many months without feeding. Using a sit-and-wait tactic while remaining in its burrow with the head or upper body extended from the entrance, an amphiuma can often capture prey. Diet includes crayfish, earthworms, fish, snails, snakes, frogs, aquatic and terrestrial insects, ground skinks, and a small amount of vegetable matter. Cottonmouths and mud snakes are considered the primary predators of Three-toed Amphiumas. Other predators probably include alligators, fish, and wading birds. Antipredator responses include biting when grabbed, squirming wildly, and releasing slime through their skin. They have been known to draw blood with their bites when handled.

Three-toed Amphiuma, Jefferson County.

Three-toed Amphiuma, Jefferson County.

REPRODUCTION Breeding occurs in shallow water from December through June, with the peak in March after heavy rains. Viable sperm may be stored for at least 6–8 months after mating. It may take Three-toed Amphiumas 3–4 years to reach sexual maturity. Eggs are deposited in rosary-like strings 50–120 mm (2.0–4.7 in.) long in a nesting cavity. The female will remain coiled around the eggs until hatching. Within about 3 weeks after hatching, the larvae lose their external gills and transformation occurs.

COMMENTS AND CONSERVATION This species is on the TPWD's Black List. Three-toed Amphiumas are locally common in many areas as long as the water is unpolluted, and they do not appear in immediate need of protection. This species has been reported to still be moderately abundant in cypress swamps and oxbows on either side of the Neches River in the Big Thicket.

FAMILY PLETHODONTIDAE: LUNGLESS SALAMANDERS

This is the largest and most successful family of living salamanders, which is made up of more than 418 species in North America, tropical America, southern Europe, and the Korean Peninsula. They are found southward from Canada through the United States, Mexico, and Central America into South America; 3 genera occur in southern Europe, and 1 genus is found in Korea. The only genera occurring in Texas are *Desmognathus*, *Eurycea*, and *Plethodon*. Characteristics of this family include the absence of lungs, 4 well-developed legs, and nasolabial grooves; sexually active males of most species have cirri, or nasal swellings, associated with the nasolabial grooves, mental glands that are used in courtship, and papillose cloacal lips. Plethodontids may be terrestrial, semiaquatic, or aquatic. Terrestrial forms usually lay their eggs in moist places on land. Semiaquatic forms live near streams and lay their eggs in the water. Aquatic forms (actually, neotenic forms) spend their lives in springs or caves. A few species of the aquatic forms have degenerate eyes and reduced skin pigmentation.

Salado Salamander
Eurycea chisholmensis,
Chippindale, Price, Wiens,
and Hillis, 2000

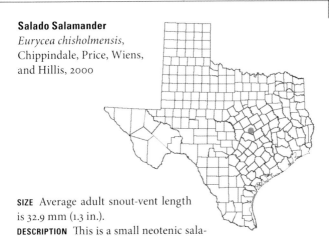

SIZE Average adult snout-vent length
is 32.9 mm (1.3 in.).

DESCRIPTION This is a small neotenic salamander with 3 pair of reddish-brown gills,
grayish-brown dorsal coloration, and a translucent belly. Legs
are cream colored, speckled with brown dorsally. A prominent
gold stripe is present on the dorsal surface of the tail, which has
well-developed dorsal and ventral caudal fins. There are 18 presacral vertebrae, with either 15 or 16 costal grooves. There are 4
digits present on the forelimbs, 5 on the hind limbs.

SIMILAR SPECIES The Salado Salamander has reduced eyes and
lacks the well-defined iridophores and dark eye rings found
in nearby spring-dwelling *Eurycea* species (*E. naufragia* and
E. tonkawae).

DISTRIBUTION The most northeastern population of *Eurycea* in
Central Texas, the Salado Salamander is restricted to 2 springs
in Salado (Bell County).

NATURAL HISTORY Little is known. The Salado Salamander is
thought to be completely aquatic and found only in the vicinity of spring outflows, either under rocks or in gravel substrate.
Captive animals have been observed feeding on amphipods.

REPRODUCTION Little is known about reproduction in this species.
Since this is a paedomorphic species, juveniles are miniature
versions of the adults. Few juveniles have been collected; a single
juvenile collected in the late 1940s was the only specimen known
from this population for decades.

COMMENTS AND CONSERVATION The restricted distribution of this
species is a primary concern for conservation. The type local-

Salado Salamander, Bell County. Photo by Kenny Wray.

ity is within a municipal park, and most spring outlets have been heavily modified over the past 150 years. Additionally, recent groundwater contamination incidents have occurred. The Salado Salamander is a candidate species for listing by the U.S. Fish and Wildlife Service (USFWS; listing priority 2), and is on the TPWD's Black List.

Cascade Caverns Salamander
Eurycea latitans, Smith and Potter, 1946

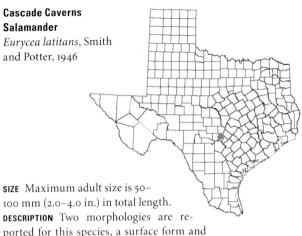

SIZE Maximum adult size is 50–100 mm (2.0–4.0 in.) in total length.

DESCRIPTION Two morphologies are reported for this species, a surface form and a cave form. The surface populations are yellow-brown above with 10–12 pairs of small dorsolateral light spots running from the head to the beginning of the tail. A dark bar is present between the eye and nostril. Eyes are normal sized, and the ventral surface of the entire body is a translucent white. Cave populations have considerably less pigmentation, especially dorsally and on the gills, and have eyes that can be greatly reduced (even beneath surface of skin); cave forms may have a more flattened snout and a sloping forehead. There are some populations that metamorphose as adults and lose their gills. There are 4 toes on the front feet, 5 on the hind feet, and 14–15 costal grooves.

SIMILAR SPECIES Many other Edwards Plateau *Eurycea* may appear remarkably similar in shape and form, and the variation seen between cave and surface populations, as well as in some surface populations undergoing complete metamorphosis, has added to the taxonomic confusion and distinctiveness of this group. There may be more than one distinct species under this single name, so at this time, geographic restrictions (see below) may be the most helpful in distinguishing between other named species and *E. latitans*.

DISTRIBUTION Originally described from Cascade Caverns in Kendall County, more recent molecular work (Chippindale et al. 2000) has included populations found in springs in Comal, Kerr, and Hays Counties as well as additional springs in Ken-

Cascade Caverns Salamander, Kendall County.

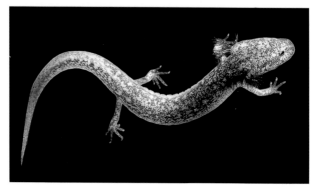

Cascade Caverns Salamander, Kendall County. Photo by Danté Fenolio.

dall County within this species complex. Future work may reveal *E. latitans* to comprise several unique species.

NATURAL HISTORY *Eurycea latitans* is strictly an aquatic salamander living in freshwater cave pools and streams. Little is known about behavior or reproduction in this species.

REPRODUCTION Unknown. Eggs are likely similar to those of other *Eurycea* observed in captivity: laid singly, clutch sizes ranging from 19 to 50, with an embryonic period lasting 2–4 weeks. Larvae spend time under cover or gravel during day. Growth is likely similar to that of *E. neotenes* in captivity.

COMMENTS AND CONSERVATION The Cascade Caverns salamander is considered a threatened species by the TPWD and is fully protected by the State of Texas.

San Marcos Salamander
Eurycea nana, Bishop, 1941

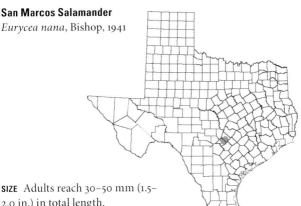

SIZE Adults reach 30–50 mm (1.5–2.0 in.) in total length.

DESCRIPTION The limbs are slender, as is the body, which has a finned tail and light-colored external gills. The dorsal surface is light brown with 7–9 small yellow spots arranged in a lateral row along each side of the back. The ventral body surface is white and translucent; the ventral surface of the tail is pale yellow. Dark rings encircle the small eyes. There are 4 toes on the front feet, 5 on hind feet, and 16–17 costal grooves.

SIMILAR SPECIES *Eurycea nana*, which is found only in and around Spring Lake (Hays County), should not be confused with any other species. *Eurycea rathbuni*, the only other species found at this locality, can be distinguished from *E. nana* by its elongated body, spatulate head, and long spindly legs.

DISTRIBUTION *Eurycea nana* is restricted to the spring outflows of Spring Lake in San Marcos (Hays County) as well as to the riffle habitat just below Spring Lake Dam.

NATURAL HISTORY Little is known. Individuals live in mats of blue-green algae that cover the streambed, and beneath rocks and gravel. Strictly aquatic, *Eurycea nana* may be seen among algae in the spring-fed pool at head of the San Marcos River. Invertebrates make up the diet of this species, especially chironomids and amphipods. Potential predators include sunfish, catfish, and crayfish.

REPRODUCTION Little is published regarding their reproduction. In captivity, eggs have been deposited on moss and filamentous algae as well as on rocks and glass marbles. An egg-laying event

San Marcos Salamander, Hays County.

San Marcos Salamander, Hays County.

for this salamander averages 33 eggs. Eggs laid in captivity hatch 16–35 days after oviposition. As in other paedormorphic species, the juveniles are small versions of the adults.

COMMENTS AND CONSERVATION The San Marcos Salamander is considered a threatened species by the TPWD and is fully protected by the state. Additional protection has been afforded by the USFWS, which considers the San Marcos Salamander a federally threatened species.

Georgetown Salamander
Eurycea naufragia,
Chippindale, Price, Wiens,
and Hillis, 2000

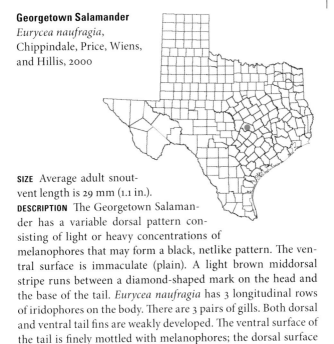

SIZE Average adult snout-vent length is 29 mm (1.1 in.).

DESCRIPTION The Georgetown Salamander has a variable dorsal pattern consisting of light or heavy concentrations of melanophores that may form a black, netlike pattern. The ventral surface is immaculate (plain). A light brown middorsal stripe runs between a diamond-shaped mark on the head and the base of the tail. *Eurycea naufragia* has 3 longitudinal rows of iridophores on the body. There are 3 pairs of gills. Both dorsal and ventral tail fins are weakly developed. The ventral surface of the tail is finely mottled with melanophores; the dorsal surface is golden yellow with concentrations of melanophores along the lateral margins of the yellowish dorsal tail fin. There are 4 digits on the forelimbs, 5 on the hind limbs, 14–16 costal grooves, and 17 presacral vertebrae.

SIMILAR SPECIES Spring-dwelling individuals of the Jollyville Salamander are most similar to Georgetown Salamanders. But Georgetown Salamanders have light starburst-shaped areas surrounding each dorsolateral iridophore, whereas most Jollyville Salamanders have square- or oblong-shaped light areas.

DISTRIBUTION This species is primarily found along the drainages of the South, Middle, and North Forks of the San Gabriel River, west of Georgetown (Williamson County). Some populations have been located within the Georgetown city limits, but these populations may be tenuous at best. A few additional populations are known in the Berry Creek drainage, which flows into the San Gabriel River east of Georgetown.

NATURAL HISTORY Little is known. The Georgetown Salamander is thought to be completely aquatic and is found only in the vi-

Georgetown Salamander, Williamson County. Photo by Kenny Wray.

cinity of spring outflows, either under rocks or in gravel substrates. Small aquatic invertebrates likely compose the diet of the Georgetown Salamander. Time of day does not influence surface counts of this species, suggesting its movements are linked to spring flow rates and predation pressure.

REPRODUCTION Information on reproduction is unknown. As in other paedomorphic species, juveniles are miniature versions of the adults.

COMMENTS AND CONSERVATION Little information is known about historical abundances for this species, since most known localities have been found only in the last 15 years. This species is a candidate for listing by the USFWS (listing priority 8), and is on the TPWD's Black List.

Texas Salamander
Eurycea neotenes, Bishop
and Wright, 1937

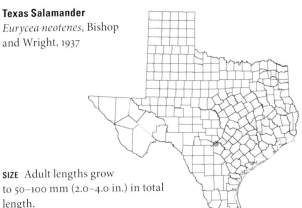

SIZE Adult lengths grow
to 50–100 mm (2.0–4.0 in.) in total
length.

DESCRIPTION The Texas Salamander has a
slender body with short limbs and bright
red external gills. The tail contains narrow dorsal and ventral
fins. The body color is light brown to yellow, with darker brown
mottling and 2 rows of lighter flecks on each side of the body. A
dark bar is present between the eye and nostril. The translucent
ventral surface is a pale cream color; the sides of head and chin
are lightly pigmented. There are 4 digits on the forelimbs, 5 on
the hind limbs, and 15–17 costal grooves.

SIMILAR SPECIES Overall morphological similarities make it extremely difficult to differentiate this species from many other
spring-dwelling *Eurycea* of the Hill Country. Recent detailed
morphological and molecular analyses have restricted this species to a handful of springs in northwestern Bexar and southern
Kendall Counties.

DISTRIBUTION Because of their morphological similarity, many
Eurycea populations from across the Edwards Plateau were pre-
viously grouped together and called *E. neotenes*; however, recent
genetic data indicates that this species is restricted to Helotes
Creek Spring, Leon Springs, and Mueller's Spring in Bexar and
Kendall Counties in south-central Texas.

NATURAL HISTORY Strictly aquatic, *Eurycea neotenes* is found in
subterranean streams and creek headwaters, remaining under
rocks and among the rock cobbles at the bottom of streambeds.

REPRODUCTION Almost nothing is known about reproduction in
this species in the wild; one courtship sequence observed in the
laboratory took almost 3 hours from start to finish. In one study,

Texas Salamander, Bexar County.

Texas Salamander, Bexar County.

gravid females were collected only in the spring and early summer. Egg oviposition has been observed only in captive *Eurycea neotenes*, with 10 eggs laid on plants, rocks, and the bottom of the cage. Small juveniles were observed in March, May, and June.

COMMENTS AND CONSERVATION The Texas Salamander is on the TPWD's Black List. This species may be considered a candidate for federal listing because of the present or threatened destruction, modification, or curtailment of its habitat or range resulting from drought. A large die-off from "red leg" disease (a bacterial infection) was observed in this species along Helotes Creek in 1958 and 1959.

Fern Bank Salamander

Eurycea pterophila, Burger,
Smith, and Potter, 1950

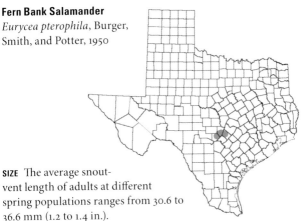

SIZE The average snout-
vent length of adults at different
spring populations ranges from 30.6 to
36.6 mm (1.2 to 1.4 in.).

DESCRIPTION The Fern Bank Salamander
has a slender body with short limbs and reddish-brown external
gills. The dorsal body color is a finely mottled brown and yellow,
darker laterally and on the dorsal surface of the head, and with
2 rows of lighter flecks on each side of the body. The dorsal tail
fin is bordered by a dull orange stripe; brown mottling extends
laterally on the tail. The ventral surfaces of the body and chin
are pale yellow. There are 4 digits on the forelimbs, 5 on the hind
limbs, and 16 costal grooves.

SIMILAR SPECIES This can be an extremely difficult species to distinguish morphologically from other Edwards Plateau *Eurycea*,
including *E. neotenes*. In fact, the type description reads, "In-
distinguishable from *Eurycea neotenes* in external morphology"
(Burger, Smith, and Potter 1950). Subtle differences in skeletal
morphology can be used to separate *E. pterophila* from other
species, but recent detailed molecular analyses have restricted
this species to the Blanco River drainage in Blanco, Hays, and
Kendall Counties.

DISTRIBUTION This species is found in and along springs in the
Blanco River drainage in Blanco, Hays, and Kendall Counties.

NATURAL HISTORY This species is known from both cave and surface populations. Spring populations are known to have sur-
vived drought conditions when springs ceased to flow. Detailed
feeding studies have not been performed. This species is thought
to feed on small aquatic invertebrates. Spring populations are

Fern Bank Salamander, Blanco County.

Fern Bank Salamander, Blanco County.

generally found immediately adjacent to the spring outflow, within a gravel substrate or under rocks and leaves.

REPRODUCTION It is thought that they deposit eggs in a gravel substrate, like other spring-dwelling *Eurycea* species. As in other paedomorphic species, juveniles are miniature versions of the adults. Juveniles are more likely than adults to be found in shallower water with smaller cobble.

COMMENTS AND CONSERVATION The Fern Bank Salamander is on the TPWD's Black List. The original description of the species listed short digits on all limbs, but this is now believed to have been related to tissue loss from a bacterial infection (*Aeromonas* sp.) in the type series of specimens.

Dwarf Salamander
Eurycea quadridigitata,
(Holbrook, 1842)

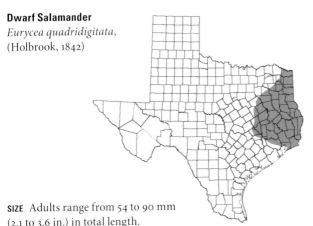

SIZE Adults range from 54 to 90 mm
(2.1 to 3.6 in.) in total length.

DESCRIPTION The dorsal coloration includes
a broad bronze to yellowish-brown stripe, and is bordered by a narrower, inconspicuous, brown to black dorsolateral stripe extending onto the tail. A row of small dark spots may be present on the dorsal surface. The ventral surface tends to be much lighter than the dorsal surface, with colors ranging from cream to yellow to bronze. The body and tail are very long and slender; the keeled tail may be as much as 1.5 times the length of the head and body. There are small, fully developed legs with 4 toes on each front and hind foot. The head is small, with well-developed, protruding eyes. During the breeding season, sexually active males have cirri and elongated monocuspid teeth. There are 14–17 costal grooves.

SIMILAR SPECIES This is the only terrestrial salamander in Texas that has 4 toes on the hind feet as well as the front feet.

DISTRIBUTION The Dwarf Salamander is found primarily in the Atlantic and Gulf Coastal Plains from North Carolina southward into peninsular Florida and westward into eastern Texas and southern Arkansas.

NATURAL HISTORY Adults have been found near swampy areas, ponds, lakes, seepages, springs, and sphagnum bogs, hiding under logs, vegetative debris, and other objects on the surface from January to April. During the summer and fall, they may be found a good distance away from water, usually under

Dwarf Salamander, San Augustine County.

Dwarf Salamander, Jasper County.

logs or deep leaf litter. While migrating to their breeding sites, many may be found crossing roads during heavy rains. The larvae are primarily benthic feeders and prey on small zooplankton, along with ostracods, cladocerans, and chironomid larvae. Adults feed on small invertebrates, including amphipods, flies, beetles and larval beetles, earthworms, hemipterans, homopterans, ants, wasps, collembolans, spiders, pseudoscorpions, mites, ticks, and millipedes. Predators may include other amphibians (especially large frogs), snakes, birds, and large invertebrates. Adults avoid predation and desiccation by traveling under the leaf litter.

REPRODUCTION Migrations to their breeding sites occur from October to February in Texas. They are more likely to be observed moving during the daylight than most other salamanders. Males have been observed in breeding condition in late fall. Females have been collected with spermatophores in their vents from September to November. Females lay 12–48 eggs singly or in small clusters attached to the undersides of submerged vegetation, such as twigs, rootlets, and debris. Larvae hatch in 30–40 days. The hind limbs are not completely developed at hatching. Transformation occurs within 2–6 months after hatching.

COMMENTS AND CONSERVATION The Dwarf Salamander is on the TPWD's Black List. These salamanders are abundant at some localities in Texas; however, in many areas throughout its range, populations have been eliminated because of loss of wetland habitat. *Eurycea quadridigitata* may represent a species complex with several different species within the group.

Texas Blind Salamander
Eurycea rathbuni,
(Stejneger, 1896)

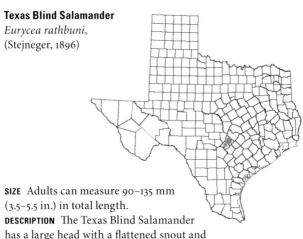

SIZE Adults can measure 90–135 mm
(3.5–5.5 in.) in total length.

DESCRIPTION The Texas Blind Salamander
has a large head with a flattened snout and
reduced, vestigial eyes beneath the surface of the skin. They are
white to pale pink with translucent skin and bright red external
gills that are retained as adults. *Eurycea rathbuni* has thin, elongated limbs; the fore- and hind limbs overlap each other when
adpressed. The compressed tail is more strongly finned dorsally than ventrally, and tapers to a point. There are 4 toes on the
front feet, 5 on the hind feet, and 12 costal grooves.

SIMILAR SPECIES No other salamanders live in the same subterranean caverns as *Eurycea rathbuni*. The San Marcos Salamander
(*E. nana*) can be found in spring habitats directly above the cav-
ern habitats of *E. rathbuni*. *Eurycea rathbuni* can be easily dis-

Texas Blind Salamander, Hays County.

tinguished from *E. nana* by its large head, flattened snout, vestigial eyes, and elongated limbs. The other 2 species of blind salamanders are found in habitats farther to the north and are distinguished by either their more robust body form and broad, rounded tail (*E. robusta*) or their lack of a well-developed tail fin and proportionally shorter legs (*E. waterlooensis*).

DISTRIBUTION The Texas Blind Salamander is restricted to Texas. It is found in the subterranean streams of the Purgatory Creek system, and is seen aboveground only when water flow brings it to the surface. *Eurycea rathbuni* is found only in the Balcones Escarpment near San Marcos (Hays County).

NATURAL HISTORY *Eurycea rathbuni* is completely aquatic and feeds on a variety of invertebrates, including shrimp, snails, and amphipods.

REPRODUCTION Little is known about reproduction in *Eurycea rathbuni* except that gravid females can be observed throughout the year. Eggs are known only from captive populations. One female was reported to have 39 mature ova. As in other paedomorphic species, juveniles are miniature versions of their adult counterparts.

COMMENTS AND CONSERVATION The Texas Blind Salamander is considered an endangered species by the TPWD and is fully protected by the state. Additional protection has been afforded by the USFWS, which considers the Texas Blind Salamander a federally endangered species. Populations of blind salamanders recently discovered in aquifers in Comal County have tentatively been designated as *Eurycea rathbuni*.

Texas Blind Salamander, Hays County.

Texas Blind Salamander, Comal County.
Photo by Danté Fenolio.

Blanco Blind Salamander
Eurycea robusta, (Longley, 1978)

SIZE The only known specimen measured 100 mm (4.0 in.) in total length.

DESCRIPTION The Blanco Blind Salamander has reduced vestigial eyes beneath the surface of the skin, and bright red external gills. *Eurycea robusta* is white to pale pink with translucent skin. The thin, elongated limbs have 4 toes on the front feet and 5 on the hind feet. They have 12 costal grooves. They have a finned tail that tapers at the tip.

SIMILAR SPECIES The only other blind cave salamanders are *Eurycea rathbuni* and *E. waterlooensis*. *Eurycea robusta* is distinguished from *E. rathbuni* by its more robust body form, a broad, rounded tail, and a rounded skull. *Eurycea robusta* can be distinguished from *E. waterlooensis* by its well-developed tail fin and proportionally longer limbs.

DISTRIBUTION The Blanco Blind Salamander has been found only in the Balcones Aquifer near the Blanco River in Hays County, Texas.

NATURAL HISTORY *Eurycea robusta* is known from 4 specimens observed in 1951 (only 1 was collected and preserved). This species was seen only this once, when workers were drilling in a streambed; it is guessed that the species inhabits subterranean streams.

REPRODUCTION Unknown

COMMENTS AND CONSERVATION The Blanco Blind Salamander is listed as a threatened species by the TPWD and is fully protected by the state. This species is also listed by the USFWS as species for which data indicates that listing as threatened or endangered may be warranted because of the present or threatened destruction, modification, or curtailment of its habitat or range resulting from water pollutants and water withdrawal.

Barton Springs Salamander

Eurycea sosorum,
Chippindale, Hillis,
and Price, 1993

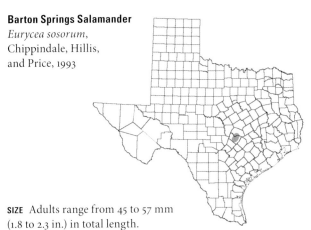

SIZE Adults range from 45 to 57 mm
(1.8 to 2.3 in.) in total length.

DESCRIPTION The Barton Springs Salaman-
der has a slender body with a flattened snout and 3 pairs of
bright red external gills. The dorsal ground color is yellowish
cream with olive-brown mottling; the ventral surface is a translucent cream white. There is a relatively short finned tail with
a narrow orange-yellow stripe. The elongated limbs have 4 toes
on the front feet and 5 on the hind feet. There are 14–15 costal
grooves.

SIMILAR SPECIES Only 2 salamander species occur at Barton
Springs (Travis County): *Eurycea sosorum* and *E. waterlooensis*.
The latter species is extremely uncommon to find away from its
preferred underground karst habitat and can be distinguished

Barton Springs Salamander, Travis County.

Barton Springs Salamander, Travis County.

by its larger size, elongated head, absence of eyes, and 12 costal grooves.

DISTRIBUTION The Barton Springs Salamander is found only in the Barton Springs Pool area in Austin.

NATURAL HISTORY *Eurycea sosorum* feeds on amphipods, earthworms, and fairy shrimp. The species is thought to be mostly a surface dweller, but it is able to live underground. *Eurycea sosorum* is strictly aquatic and may be found among rubble in the spring outflow at Barton Springs.

REPRODUCTION Little is known about reproduction in *E. sosorum*, but recently hatched young have been found in the months of November, March, and April. Reports about eggs and oviposition come from captive breeding programs. Females have laid as many as 29 eggs, which hatch in 25–35 days. Oviposition takes place on cobble, aquatic plants, and the sides of the aquaria. As in other paedomorphic species, juveniles are miniature versions of adults.

COMMENTS AND CONSERVATION The Barton Springs Salamander is considered an endangered species by the TPWD and is fully protected by the state. Additional protection has been afforded by the USFWS, which considers the Barton Springs Salamander a federally endangered species.

Jollyville Plateau Salamander

Eurycea tonkawae,
Chippindale, Price, Wiens,
and Hillis, 2000

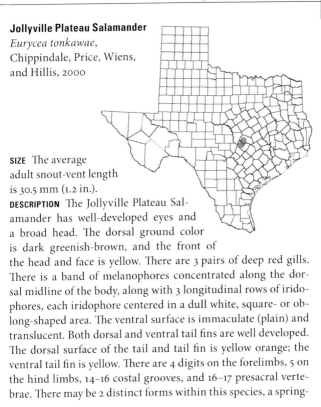

SIZE The average
adult snout-vent length
is 30.5 mm (1.2 in.).

DESCRIPTION The Jollyville Plateau Salamander has well-developed eyes and
a broad head. The dorsal ground color
is dark greenish-brown, and the front of
the head and face is yellow. There are 3 pairs of deep red gills.
There is a band of melanophores concentrated along the dorsal midline of the body, along with 3 longitudinal rows of iridophores, each iridophore centered in a dull white, square- or oblong-shaped area. The ventral surface is immaculate (plain) and
translucent. Both dorsal and ventral tail fins are well developed.
The dorsal surface of the tail and tail fin is yellow orange; the
ventral tail fin is yellow. There are 4 digits on the forelimbs, 5 on
the hind limbs, 14–16 costal grooves, and 16–17 presacral vertebrae. There may be 2 distinct forms within this species, a spring-

Jollyville Plateau Salamander, Travis County.

Jollyville Plateau Salamander, Travis County.

dwelling form and a cave-dwelling form, but they are distinguishable only at the allozyme level.

SIMILAR SPECIES Spring-dwelling individuals of the Jollyville Plateau Salamander are most similar to Georgetown Salamanders (*Eurycea naufragia*). But Georgetown Salamanders have light starburst-shaped areas surrounding each dorsolateral iridophore, whereas most Jollyville Plateau Salamanders have square- or oblong-shaped light areas.

DISTRIBUTION The Jollyville Plateau Salamander is found at a number of springs along creeks in western Travis and Williamson Counties. These creeks include Brushy Creek, Bull Creek, Shoal Creek, and Walnut Creek.

NATURAL HISTORY Little is known. *Eurycea tonkawae* is thought to be completely aquatic and is found only in the vicinity of spring outflows, either under rocks or in gravel substrates. Salamander abundance is positively correlated with suitable areas of cobble, but negatively correlated with urbanization. Spring populations likely make extensive use of subterranean habitats during periods of drought; adult salamanders have been found at sites where spring flow was restored after being dry for months. Small aquatic invertebrates likely compose the diet of the Jollyville Plateau Salamander. Prey includes small aquatic invertebrates such as copepods, ostracods, chironomid larvae, and physid snails. Sunfish and crayfish may be significant predators of spring populations.

REPRODUCTION It is thought that they deposit eggs in gravel substrates, like other spring-dwelling *Eurycea* species. As in other paedomorphic species, juveniles are miniature versions of the adults. Juveniles are more likely to be found in shallower water with smaller cobble than adults.

COMMENTS AND CONSERVATION Little information is known about historical abundances for this species, since most known localities have been found only in the last 15 years. Some historical localities have been destroyed by quarry operations or building construction. Most localities are located in northwest Austin, where continued development is fueled by a burgeoning human population. Many populations are threatened by habitat degradation and destruction. This species is a candidate for listing as endangered or threatened by the USFWS (listing priority 8), and is on the TPWD's Black List.

Comal Blind Salamander
Eurycea tridentifera,
Mitchell and Reddell, 1965

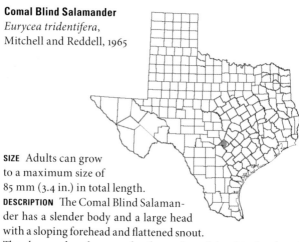

SIZE Adults can grow
to a maximum size of
85 mm (3.4 in.) in total length.

DESCRIPTION The Comal Blind Salamander has a slender body and a large head with a sloping forehead and flattened snout. They have reduced eyes under the surface of the skin, bright red external gills, thinly elongated limbs, and a finned tail. Dorsal coloration is cream or pale yellow with gray, tan, or orange mottling. Pairs (4–15) of dorsolateral light spots are found from behind the head to base of the tail. The ventral surface is immaculate and translucent. There are 4 toes on the front limbs, 5 on the hind limbs, and 11–12 costal grooves.

SIMILAR SPECIES Superficially, *Eurycea tridentifera* could be confused with *E. rathbuni*, but the latter has an extremely flattened

Comal Blind Salamander, Comal County. Photo by Joe Furman.

Comal Blind Salamander, Comal County. Photo by Danté Fenolio.

snout and exceptionally long legs for any *Eurycea* species. The 12 costal grooves of *E. tridentifera* are fewer than those in most other *Eurycea* species, and the lack of well-developed eyes make this species easy to distinguish from all other surface populations of *Eurycea*.

DISTRIBUTION The Comal Blind Salamander is known to occur in the southeastern margin of the Edwards Plateau and the Cibolo Sinkhole Plain region of Comal and Bexar Counties, and probably in caves of southeastern Kendall County.

NATURAL HISTORY *Eurycea tridentifera* is strictly aquatic and is found in the underground waters of limestone caves. This species likely feeds on small aquatic invertebrates; some researchers hypothesize that they may eat bat guano and the insects parts contained therein. When disturbed, *E. tridentifera* may swim upward rather than down toward the substrate. Hybridization with congeners may occur near cave entrances.

REPRODUCTION Little is known for this species in the wild. Clutch sizes range from 7 to 18 eggs. Metamorphosis does not occur in this species (paedomorphic), so animals retain their gills as adults.

COMMENTS AND CONSERVATION The Comal Blind Salamander is considered a threatened species by the TPWD and is fully protected by the state. This species is currently listed by the USFWS as a species for which substantial information indicates that listing as threatened or endangered may be warranted. Threats to this species and others in the Comal Springs ecosystem are habitat loss and modification because of groundwater contamination or groundwater withdrawal.

Valdina Farms Salamander
Eurycea troglodytes, Baker, 1957

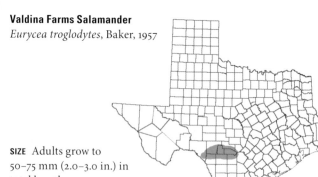

SIZE Adults grow to 50–75 mm (2.0–3.0 in.) in total length.

DESCRIPTION Two morphologies are reported for this species, a surface form and a cave form. The surface populations are light brown to yellow with darker brown mottling and 2 rows of lighter flecks on each side of the body. These rows of flecks may be organized as pale yellow stripes present laterally on the body and sometimes on the dorsal surface of the tail. A dark bar is present between each eye and nostril. Eyes are normal sized, and the ventral surface of the entire body is a translucent white. They have bright red external gills. Cave populations have considerably less pigmentation, especially dorsally and on the gills, and have eyes that may be greatly reduced (even beneath surface of skin); cave forms may have a more flattened snout and a sloping forehead. There are some populations that metamorphose as adults and lose their gills. There are 4 toes on the front limbs, 5 on the hind limbs, and 13–14 costal grooves.

Valdina Farms Salamander, Edwards County.

Valdina Farms Salamander, Edwards County.

SIMILAR SPECIES Many other Edwards Plateau *Eurycea* may appear remarkably similar in shape and form, and the variation seen between cave and surface populations, as well as in some surface populations undergoing complete metamorphosis, has added to the taxonomic confusion and distinctiveness of this group. There may be more than 1 distinct species under this single name, so at this time, geographic restrictions (see below) may be the most helpful in distinguishing between other named species and *E. troglodytes*.

DISTRIBUTION Although the species was originally described from a single locality, the Valdina Farms Sinkhole in Medina County, more recent molecular work (Chippindale et al. 2000) has included populations found in springs and caves in Bandera, Gillespie, western Kerr, Real, Uvalde, and Val Verde Counties, as well as in additional springs in Medina County, within this species complex.

NATURAL HISTORY Strictly aquatic, *Eurycea troglodytes* is found in springs and cave waters. This species is thought to feed upon small aquatic invertebrates, but some terrestrial invertebrates have been found in the stomachs of transformed individuals. During the day, this species seeks shelter in gravel substrates or under the cover of aquatic vegetation.

REPRODUCTION Unknown, but reproduction and egg laying are thought to be similar to that of other Edwards Plateau *Eurycea* (eggs laid in a gravel substrate). The species is generally thought to be paedomorphic, but complete metamorphosis has been ob-

109

Valdina Farms Salamander, transformed adult, Bandera County.

served in populations along the Sabinal River drainage (Bandera County) as well as in a cave in Uvalde County.

COMMENTS AND CONSERVATION The Valdina Farms Salamander is on the TPWD's Black List. In the 1980s, the type locality (Valdina Farms Sinkhole) was altered to allow floodwaters from nearby Seco Creek to fill the sinkhole so that it could better serve as an aquifer-recharge structure. The effects of this project on salamander populations are likely twofold. First, it allowed for the introduction of catfish and other predators into the sinkhole. Second, it eliminated the sinkhole's roosting colony of Mexican Free-tailed Bats, whose guano served as the base of the sinkhole's food chain. These 2 effects have likely decimated the local population of these salamanders, since none have been located at this locality in surveys since the completion of the project. Future work may reveal *Eurycea troglodytes* to comprise several unique species.

Austin Blind Salamander
Eurycea waterlooensis,
Hillis, Chamberlain,
Wilcox, and Chippendale,
2001

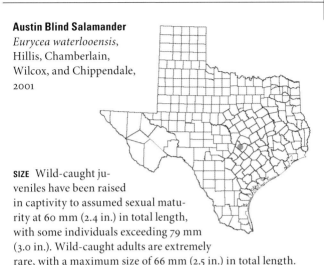

SIZE Wild-caught juveniles have been raised
in captivity to assumed sexual maturity at 60 mm (2.4 in.) in total length,
with some individuals exceeding 79 mm
(3.0 in.). Wild-caught adults are extremely
rare, with a maximum size of 66 mm (2.5 in.) in total length.

DESCRIPTION The Austin Blind Salamander has an elongated
head, with the anterior skull laterally expanded to produce an
extended snout. This species lacks external eyes; dark eyespots
present under the skin. Gills are present. A translucent skin gives
the body a pearly white appearance; light lavender coloration
dorsally is caused by a weak but uniform distribution of melanophores. Some specimens contain a lateral row of iridophores
that extends from the body onto the tail. Tail fins are weakly developed (the ventral portion is present only on the anterior half

Austin Blind Salamander, Travis County.

Austin Blind Salamander, Travis County.

of the tail). There are 4 digits on the forelimbs, 5 on the hind limbs, 12 costal grooves (the anterior-most and posterior-most being weakly defined), and 15 presacral vertebrae.

SIMILAR SPECIES The Blanco Blind Salamander (*Eurycea robusta*) and the Texas Blind Salamander (*E. rathbuni*) are similar-looking salamanders with flattened heads and barely visible, dark eyespots. In addition to its restricted, nonoverlapping distribution, the Austin Blind Salamander is shorter in length, has more dorsal pigmentation, lacks a well-developed tail fin, and has proportionally shorter legs than the other 2 blind salamanders. The Barton Springs Salamander has well-defined eyes and a normal snout in contrast to the dark eyespots and flattened snout of the Austin Blind Salamander.

DISTRIBUTION The Austin Blind Salamander is known from only 3 outflows at Barton Springs (Travis County). Along with the Barton Springs Salamander, this is 1 of 2 salamander species restricted to the Barton Springs complex.

NATURAL HISTORY The Austin Blind Salamander is entirely aquatic, and is probably restricted to subterranean features of the Edwards Aquifer, with juveniles (and adults less frequently) accidentally being flushed to the surface periodically. This salamander feeds on a variety of small aquatic invertebrates in captivity;

wild-caught individuals have defecated remnants of amphipods, copepods, and plant material. This species is thought to be active year-round, with the majority of observations occurring from November to January.

REPRODUCTION Unknown. Most individuals observed in springs are juveniles. Individuals have been raised in captivity from juveniles to adults in 8 months.

COMMENTS AND CONSERVATION The City of Austin currently has a captive propagation facility that maintains a population of this species. The Austin Blind Salamander is on the TPWD's Black List and is a candidate species for listing by the USFWS (listing priority 2).

Comal Springs Salamander

Eurycea species #1
(New Braunfels)

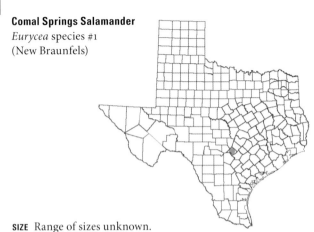

SIZE Range of sizes unknown.

DESCRIPTION Based on morphological and molecular analyses, Chippindale et al. (1998) and Chippindale et al. (2000) recognized this population as potentially representing a unique taxon, but no formal detailed description has been published.

SIMILAR SPECIES The San Marcos Salamander (*Eurycea nana*) and the Texas Salamander (*E. neotenes*) are the species most often confused with the Comal Springs population. But this population appears to be restricted to Comal Springs.

Comal Springs Salamander, Comal County.

Comal Springs Salamander, Comal County.

DISTRIBUTION This taxon is restricted to Comal Springs in Landa Park and Landa Lake, at the head of the Comal River in New Braunfels (Comal County).

NATURAL HISTORY Unknown. They probably feed on small aquatic invertebrates.

REPRODUCTION Unknown for wild populations. This salamander has been successfully bred in captivity in large acrylic columns packed with large gravel through which water was pumped. Females traveled down through the columns to deposit their eggs. Clutch sizes in captivity were 19–50. Captive-raised animals took 6 months to reach sexual maturity.

COMMENTS AND CONSERVATION This population of salamanders is not specifically protected in Texas. It is, however, currently listed by USFWS as a taxon ("Comal Springs salamander, species 8") for which substantial information indicates that listing as threatened or endangered may be warranted. Threats to this species and others in the Comal Springs ecosystem include habitat loss and modification from groundwater contamination or groundwater withdrawal within the Edwards Aquifer. A captive population is maintained at a USFWS facility in San Marcos.

Pedernales River Springs Salamander
Eurycea species #2

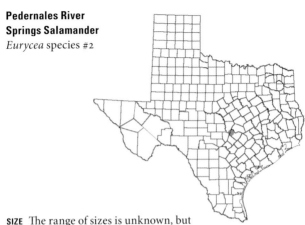

SIZE The range of sizes is unknown, but the species is thought to be smaller than most Edwards Plateau *Eurycea*.

DESCRIPTION Based on unique molecular results for these populations, Chippindale et al. (2000) recognized these populations as a unique taxon, but no formal detailed description has been published. Many individuals in the population exhibit a light yellowish-gold color created by widely spaced melanophores.

SIMILAR SPECIES This undescribed species could be confused with proximate species of *Eurycea* (for example, *E. tonkawae*), but this taxon is thought to be restricted to 2 springs in western Travis County.

DISTRIBUTION This taxon is restricted to the western edge of Travis County, along the northeast side of the Pedernales River.

NATURAL HISTORY Unknown. This salamander likely feeds on small aquatic invertebrates while living in gravel substrates in the vicinity of spring outflows.

REPRODUCTION Details are unknown, but this species is thought to mature at a very small size. As in other paedomorphic species, there is no metamorphosis, and adults retain their gills.

COMMENTS AND CONSERVATION This population of salamanders is not specifically protected in Texas. These populations could be considered a candidate for federal listing because of the present or threatened destruction, modification, or curtailment of their habitats or ranges resulting from drought and continued development of residential or commercial properties.

Southern Dusky Salamander
Desmognathus auriculatus,
(Holbrook, 1838)

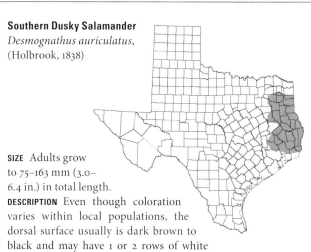

SIZE Adults grow
to 75–163 mm (3.0–
6.4 in.) in total length.

DESCRIPTION Even though coloration
varies within local populations, the
dorsal surface usually is dark brown to
black and may have 1 or 2 rows of white
to reddish spots along each side of the body between the front
and back legs; these spots may continue onto the tail. The ventral surface is grayish-brown to black and usually has fine white
specks. Characteristics of adults include a stout body, a head tapering to the snout, a groove that passes from each nostril downward to the mouth, a laterally compressed and keeled tail, hind
legs larger than front legs, toe tips lacking cornifications, and an
average of 14 costal grooves. Like other plethodontid salamanders, this species is lungless, breathing through its skin, the lin-

Southern Dusky Salamander, Tyler County.

Southern Dusky Salamander, Tyler County.

ing of its mouth, and the lining of its cloaca. Juveniles feature 6–7 pairs of dorsal spots between the limbs, and the spots may fuse to form a jagged middorsal line that disappears with age.

SIMILAR SPECIES This species could possibly be confused with the Small-mouthed Salamander (*Ambystoma texanum*) or the Western Slimy Salamander (*Plethodon albagula*). Both of these have tails that are longer than the head and body. *Ambystoma texanum* has a tail that is laterally compressed near the tip, while the tail of *D. auriculatus* is laterally compressed from the base to the tip.

DISTRIBUTION The Southern Dusky Salamander is found on the Gulf Coastal Plain in the southeastern United States from extreme southeastern Virginia south to mid-peninsular Florida, and west through southern Alabama, Mississippi, Louisiana, and southeastern Texas. In Texas, *Desmognathus auriculatus* inhabits the eastern part of the state, from the border of Oklahoma and Arkansas, south to the Gulf of Mexico.

NATURAL HISTORY These salamanders prefer murky, stagnant bodies of water with acidic soils within or near springs, cypress ponds and swamps, seeps, sloughs, bottomland habitats, floodplains, small tributaries, and slow-moving muddy streams. The Southern Dusky Salamander is nocturnal, and during the daytime it remains under leaf litter, rotten logs, peat moss, or mats of marginal aquatic vegetation, or buried in the mud. Diet includes insects, earthworms, beetle larvae, flies, small spiders, and crane flies. Water snakes, pickerel, and possibly feral pigs feed on this salamander. Cannibalism has been reported. When threatened, they may leap several times, thrash about, or simply dive into the black organic soils where they are found.

REPRODUCTION Reproduction is terrestrial. After a short courtship, males deposit a spermatophore on the substrate in front of the female. The female picks up the spermatophore with her cloacal lips. Since females have been found guarding late-term embryos or hatchlings in late August to October, it is likely that breeding takes place from late summer to early fall. Grapelike clusters of eggs are laid in cavities or small moist depressions near water. Some eggs may be placed individually on roots or other structures near water sources. Each clutch may have 9–26 eggs. Nests are within 1–2 m (3.28–6.56 ft.) of the water and have been found in sphagnum moss, in cypress logs and stumps, and beneath logs and bark. After hatching, the larvae move into the nearby water and remain in the larval stage until the following spring.

COMMENTS AND CONSERVATION This species is on the TPWD's Black List. This species seems to be in decline in Texas. Until recently, the only records of this species in the last 20 years were from Tyler County, even though efforts were made to find this salamander at many of its historical population sites. In the spring of 2011, a population of *Desmognathus auriculatus* was discovered in northern Newton County. To find out how widespread this decline in Texas may be, it is imperative that surveys and studies be conducted.

Southern Dusky Salamander, Newton County.

Western Slimy Salamander
Plethodon albagula,
Grobman, 1944

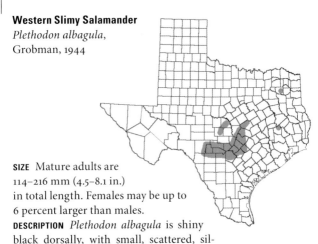

SIZE Mature adults are 114–216 mm (4.5–8.1 in.) in total length. Females may be up to 6 percent larger than males.

DESCRIPTION *Plethodon albagula* is shiny black dorsally, with small, scattered, silvery to white spots on the head, back, and tail. Larger white to yellow spots are found on the lateral sides. The ventral surface is slate to grayish black. Some populations have a light or white throat, while others have a dark or black throat. There is also a population in Texas that is almost completely black. The Western Slimy Salamander is lungless and has a body that is long and slender with 4 well-developed legs and a rounded prehensile tail that is about equal in length to the body. They have large, prominent eyes. During breeding season, males have prominent circular mental glands and small round yellow to orange glands on the belly. Males also have papillose cloacal linings. There is an average of 16 costal grooves.

VOICE When in distress, they may emit a squeak.

SIMILAR SPECIES This species might be confused with smaller *Ambystoma* specimens, but *Plethodon albagula* has a groove extending from the nostril to the lip, while *Ambystoma* lacks this groove.

DISTRIBUTION The Western Slimy Salamander is a member of a large species complex that includes several species once considered to be *Plethodon glutinosus*. Western Slimy Salamanders are found in the interior highlands of Missouri south through Arkansas and southeastern Oklahoma; a large disjunct population is found along the Edwards Plateau of Central Texas, from Bell County southeast to Edwards County. Two other disjunct populations may exist in northeast and southeast Texas. There is doubt concerning the veracity of these 2 populations, though a handful of museum specimens indicate their presence in these locations.

NATURAL HISTORY The Western Slimy Salamander may be found under logs or rocks, in cracks, and at cave entrances within wooded canyons, damp ravines, and wooded karst hillsides throughout their range. In Texas, surface nocturnal activity may occur during cool rainy periods from late fall to early spring. The diet includes flies, ants, earthworms, centipedes, millipedes, sow bugs, spiders, beetles, and isopods. Copperheads and garter snakes have been observed feeding on this species. When caught or handled, these salamanders engage in body flipping and tail lashing, which produce noxious skin secretions that are difficult to remove or wash off. Secretions that have dried onto one's hands can be peeled off. The salamanders are agile, and when disturbed, they are able to escape quickly into their underground retreats.

REPRODUCTION Terrestrial breeding takes place in late summer or fall in their underground retreats. After a short courtship, a male deposits a spermatophore in front of a female, who picks it up with her cloacal lips. Clusters of 6–36 eggs are laid in early

Western Slimy Salamander, Bandera County.

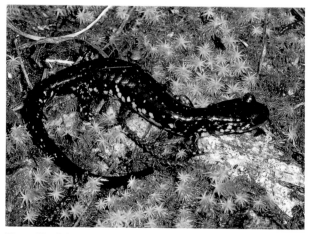

Western Slimy Salamander, Edwards County.

fall to late winter; the clusters are attached to damp vegetative debris or suspended from the walls or ceiling of caves, or placed in crevices, under large rocks, or in damp ravines. Females often guard and brood the eggs by coiling about them until they hatch. Direct development occurs after eggs are laid. The hatching of fully developed larvae occurs during winter, when cooler and damper conditions are more prevalent. Sexual maturity is reached in 3 years.

COMMENTS AND CONSERVATION The Western Slimy Salamander is on the TPWD's Black List. Although there is little natural-history information about Western Slimy Salamanders, they may be found in good numbers within their small, isolated microhabitats. Recent research indicates both morphological and molecular differentiation within Texas populations as well as between interior highland populations of Oklahoma, Arkansas, and Missouri.

FAMILY PROTEIDAE: WATERDOGS, OR MUDPUPPIES

The family Proteidae is a small family of strictly aquatic perennibranch salamanders that retain an advanced larval form throughout their lives. There are 2 genera, *Necturus* and *Proteus*, which contain 6 species. This is a widespread group found in the Old World and New World. *Proteus* is a cave form found in southern Europe. *Necturus* is restricted to eastern North America. A single species of *Necturus* is found in East Texas.

The proteids are relatively large salamanders that may reach 300–400 mm in total length. The *Necturus* salamanders are paedomorphic inhabitants of lakes and large streams; they lack maxilla, and the prevomerine teeth are in lateral rows and parallel the premaxillary teeth. Adults possess lungs, but are larviform with caudal tail fins and 3 pairs of conspicuously bushy external gills with 2 slits per side. There are 4 toes on each rear foot. Costal grooves are present, but the species lack nasolabial grooves, and the prearticular bone is fused. Fertilization is internal, and little is known about courtship.

Gulf Coast Waterdog
Necturus beyeri, Viosca, 1937

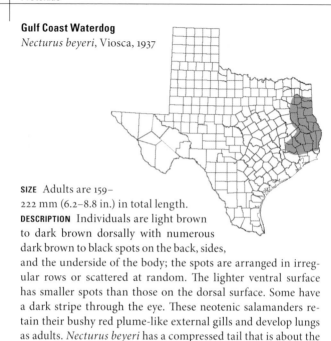

SIZE Adults are 159–222 mm (6.2–8.8 in.) in total length.

DESCRIPTION Individuals are light brown to dark brown dorsally with numerous dark brown to black spots on the back, sides, and the underside of the body; the spots are arranged in irregular rows or scattered at random. The lighter ventral surface has smaller spots than those on the dorsal surface. Some have a dark stripe through the eye. These neotenic salamanders retain their bushy red plume-like external gills and develop lungs as adults. *Necturus beyeri* has a compressed tail that is about the same length as the body, and 4 small legs with 4 toes on each

Gulf Coast Waterdog, Houston County.

Gulf Coast Waterdog, Houston County.

limb. During the breeding season, males have swollen cloacae and 2 enlarged posterior projecting cloacal papillae.

SIMILAR SPECIES Gulf Coast Waterdogs may be confused with the larvae of Mole Salamanders (*Ambystoma* species), but all *Ambystoma* larvae have 5 toes on each hind foot.

DISTRIBUTION The Gulf Coast Waterdog occurs as 2 isolated groups, the western group and the eastern group. The eastern group is found from southeastern Louisiana to central Mississippi. The western group is found in central East Texas eastward into central western Louisiana. In Texas, they are primarily found in river drainages of the San Jacinto, Neches, and Sabine.

NATURAL HISTORY These salamanders are associated with permanent aquatic habitats. They can be encountered in medium to large, slow-moving, sandy-bottomed, spring-fed streams with bottom debris, logjams, and leaf-litter beds. In Texas, Gulf Coast Waterdogs are nocturnal and most active from late fall to early spring. They spend the day in burrows in the banks or buried in debris on streambeds. In summer or when the food supply is limited, they may aestivate in burrows. Gulf Coast Waterdogs feed on crayfish, mayflies, caddis flies, dragonflies, beetles, midges, and sphaeriids. Crayfish seems to be the preferred diet. Predators have not been documented, but may include water snakes, mud snakes, large fish, and crayfish. The life span in nature is 6–7 years.

REPRODUCTION Mating occurs from late fall to early winter. Females have been collected with spermatophores in late December. Sperm may be stored for as long as 6 months. Males have swollen cloacae and sperm in December and January. In late April and May, eggs are deposited singly on the stream bottom, on the undersides of large boards, on railroad ties, on rocks, and on pine logs embedded in the sandy sections of a stream. Females have been found brooding nests in late spring. Since this species is neotenic, no abrupt metamorphosis occurs.

COMMENTS AND CONSERVATION This species is on the TPWD's Black List. Declines in populations have occurred because of siltation and pollution; perhaps this should be a species of concern in Texas.

FAMILY SALAMANDRIDAE: NEWTS

Salamandridae are primarily Eurasian, found in Europe, extreme northern Africa, and eastern Asia; only 2 genera occur in North America. They are often referred to as "typical salamanders and newts."

The Salamandridae have a rougher skin texture than most salamanders and are not slimy. Unlike other salamanders, they are relatively easy to capture and handle. The skin often appears grainy or warty, in contrast to the smooth skin of other salamanders. They lack nasolabial grooves, although costal grooves may be present but indistinct. They have 2 longitudinal rows of vomerine teeth, 1 on either side of the parasphenoid, on the palate that extends backward between the orbits. Mating and ovipositing usually occur in ponds or streams, and sexual dimorphism is exhibited in breeding coloration and the size of the tail fin. Most are relatively small, less than 120 mm (4.7 in.) in total length, but may reach to 200 mm (7.9 in.).

Black-spotted Newt

Notophthalmus meridionalis,
(Cope, 1880)

SIZE Mature adults reach 54–110 mm (2.1–4.3 in.) in total length.

DESCRIPTION The Black-spotted Newt has an olive-green dorsal coloration with conspicuous black spots on both the dorsal and ventral surfaces. The sides may be light blue-green. Yellow or gold spots may be found along either side of the back and may form an irregular line from the head on to the tail. In some specimens, there may be a faint stripe (brown to ruddy) down the center of the back. The ventral coloration is orange to yellow-orange. This stocky salamander has a large head in relation to the body, and the tail is vertically

Black-spotted Newt, adult, Hidalgo County.

Black-spotted Newt, eft, Cameron County.

compressed, with a keel or fin extending its length. Breeding males have hedonic pits in a line behind each eye, and cornified toe tips. The eft stage is not well defined. The eft is orange-red with red stripes like those of the adult.

SIMILAR SPECIES The Eastern Newt (*Notophthalmus viridescens*) has red spots, smaller dark spots, and no stripes. In breeding males, the Eastern Newt has cornified ridges on the underside of the thighs and a yellow spot on the posterior margin of the vent.

DISTRIBUTION The Black-spotted Newt ranges from southeastern Texas, south of the San Antonio River, along the Coastal Plain, and southward into northern Mexico.

NATURAL HISTORY Little is known of the life history of this species. Although this salamander is found in xeric areas, it is closely associated with shallow water in seasonally ephemeral and permanent streams, lagoons, ditches, and swampy areas containing large amounts of vegetation. They have been found in pools more than 2 m (6.6 ft.) deep. Even when the pond habitats dry up, individuals do not normally wander very far, moving to the ground surface under debris, logs, or rocks. Leeches, insects, worms, mollusks, crustaceans, and small amphibians and their eggs are included in the diets of Black-spotted Newts. Predator and defense documentation is limited, but this salamander has a skin gland that secretes a toxic substance that makes it unappealing.

REPRODUCTION Mating can occur in any month of the year, depending on rainfall patterns. A female collected containing oviductal eggs and gill rudiments suggest that gilled adults might occur in some populations. Up to 300 eggs have been found in shallow water and attached singly to submerged vegetation in shallow pools in March and April. Eggs hatch in 3–4 weeks. Juveniles remain in water until they become mature, unless pond drying or high temperatures drive them onto land.

SUBSPECIES The subspecies that occurs in Texas is the Texas Black-spotted Newt, *Notophthalmus meridionalis meridionalis* (Cope, 1880).

COMMENTS AND CONSERVATION The Black-spotted Newt is listed as a threatened species in Texas. This species has generated conservation concerns within its range because of pesticide and herbicide use as well as human-induced modification of breeding habitats.

Eastern Newt

Notophthalmus viridescens,
(Rafinesque, 1820)

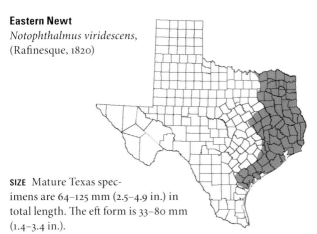

SIZE Mature Texas specimens are 64–125 mm (2.5–4.9 in.) in total length. The eft form is 33–80 mm (1.4–3.4 in.).

DESCRIPTION The adult is dorsally yellowish brown, olive green, to dark brown, and ventrally light yellow, which contrasts with the dorsal color. Small black spots may appear on the back and on the belly. Red spots may occur on the back or sides. Efts vary from bright vermilion to dull red or greenish brown, and younger individuals are more brightly colored than older ones. Red and black spotting is similar to that of the adults. These salamanders have a small head relative to the thick body. The hind legs of the adult are much larger than the front legs. During the mating season, males develop keeled tails, swollen cloacae, dark cornified toe tips, and horny black ridges and pads on the inner surfaces of the thighs. A line of hedonic pits is found in males behind the eye, and a yellowish glandular spot occurs on the posterior margins of the vent. Efts have con-

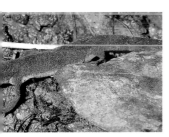

Eastern newt, eft, Brazoria County.

Eastern Newt, adult, Houston County.

spicuously granular, coarse skin. Aquatic juveniles are colored much like aquatic adults. They have laterally compressed tails, olive coloration, and smoother skin.

SIMILAR SPECIES This newt does not look like any other salamander in the state.

DISTRIBUTION The Eastern Newt is found throughout eastern United States and southeastern Canada, southward to southern Florida and westward to eastern Texas.

NATURAL HISTORY In most populations, there are 4 stages in the life history, which include the egg, aquatic larvae, terrestrial eft, and aquatic adult. There may be gilled aquatic adults or lunged adults. Adult newts reside in shallow permanent or semipermanent water habitats in forest ponds, lakes, quiet streams, ditches, farm ponds, marshes, canals, river bottoms, and swamps with dense emergent or submergent vegetation. Efts can be found under debris or logs, or in the open on the forest floor. Even thought newts are classified as aquatic salamanders, they often desert their breeding sites and reside on land, hiding under logs or debris, or in thick vegetation. The larvae transform into the terrestrial eft stage. After 1–3 years, efts migrate back to breeding sites and transform into aquatic lunged adults. They remain in the water for the remainder of their lives unless conditions force them to leave the water and live temporarily on land. Efts and terrestrial adults are most active during or after rain showers, moving about on the forest floor. Larvae hide in bottom debris or dense vegetation during the day and forage at night in the water column. *Notophthalmus viridescens* is a generalist carnivore that feeds on worms, fly larvae, water mites, mosquitoes, caddis flies, stoneflies, lepidopterans, hemipterans, beetles,

Eastern Newt, eft, Anderson County.

Eastern Newt, adult, Houston County.

dipterans, amphibian larvae and eggs, spiders, leeches, amphipods, clams, mollusks, turbellarians, cladocerans, worms, and small fish and their eggs. Both chemical and visual cues are used to locate food. Because of their toxic skin secretions, they have few predators. Efts have a greater concentration of secretions than the other stages. The red color of the efts seems to be a warning sign for most predators. Predators include some water snakes and garter snakes, hog-nosed snakes, American Bullfrogs, painted turtles, and snapping turtles. Larvae and aquatic adults may be fed upon by diving beetles, sirens, and fish. Antipredator posturing by efts includes bending the head and tail upward, closing the eyes, and curling the tail; this is known as the unken reflex.

REPRODUCTION Texas populations breed in late autumn to early spring. After an extended courtship, the male deposits 1 or more spermatophores in front of the female. She picks them up with her cloacal lips. Eggs are laid singly and are often concealed in submerged vegetation, decaying leaves, and other detritus. Eggs are laid over a period of days, scattered throughout the breeding habitat. Larvae hatch in 3–8 weeks.

SUBSPECIES The subspecies found in Texas is the Central Newt, *Notophthalmus viridescens louisianensis* Wolterstorff, 1914.

COMMENTS AND CONSERVATION This species is on the TPWD's Black List. The Eastern Newt has benefited from the construction of farm ponds. On the other hand, they have declined in recent years, in part because of the introduction of the imported fire ant. Deforestation and the drainage of ponds and marshes could endanger this newt in developing areas.

FAMILY SIRENIDAE: SIRENS

The family Sirenidae is a small group confined to North America and consisting of only 2 genera, *Pseudobranchus* and *Siren*, with 4 species, which are restricted to the eastern United States and northeastern Mexico. In Texas, *Siren* is restricted to the length of the Gulf Coastal Plain. *Pseudobranchus* is not found in Texas.

Adult sirens are long, eellike, slender salamanders with external lacy gills. These animals can reach nearly 1 m (3 ft.) in length. There are 2 greatly reduced front legs and no rear legs; a horny beak is present in the place of premaxillary teeth. There are 4 toes per limb in *Siren* salamanders, and those in *Pseudobranchus* have 3 toes per limb. These are neotenic salamanders and exhibit larval traits such as the absence of eyelids and the presence of gill slits with external gills. Females lack spermathecae, and males lack cloacal glands. Males do not produce spermatophores, and fertilization is presumed to be external.

These nocturnal salamanders lay eggs that lack pigment, and are deposited either in nests or are hidden; the eggs are attended by at least one adult at all times. Although these amphibians retain 1 or 3 pairs of gills, they have lungs and can breathe at the surface if the oxygen levels become reduced in their aquatic environment.

Lesser Siren
Siren intermedia, Barnes, 1826

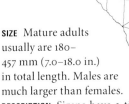

SIZE Mature adults usually are 180–457 mm (7.0–18.0 in.) in total length. Males are much larger than females.

DESCRIPTION Sirens have 2 tiny front legs, each with 4 toes. No hind legs or internal pelvic girdles are present. The tail is laterally compressed, with a fin along the top and bottom edge, and pointed at the tip. These amphibians are long and slender, generally eellike in form, with 31–36 costal grooves. The eyes are fairly well developed and have no eyelids. Sirens have external gills with 3 gill slits. Coloration is dark brown, bluish black, or olive green, with scattered black dots on the head, back, and sides. The young have a red band across the snout and along the side of the body.

VOICE Sirens are unusual among salamanders in that they make sounds consisting of a series of faint clicks when other sirens approach or when gulping air. They also sometimes emit shrill cries of distress when attacked by predators. This shrill cry is apparently the origin of the name.

SIMILAR SPECIES Within its range, the Lesser Siren can be confused only with the Rio Grande Siren. The Rio Grande Siren has 36–38 costal grooves, while the Lesser Siren has 34–36. A large adult with 36 costal grooves would likely be a Rio Grande Siren, since Rio Grande Sirens are much larger than Lesser Sirens on average. They can be distinguished from eels and amphiumas by having only 2 anterior legs with 4 toes on each leg, and distinct exterior gills; freshwater eels have no legs and no frilly gills.

DISTRIBUTION The Lesser Siren is found in the southeastern United States, ranging from the Atlantic Coastal Plain of southeastern North Carolina to central Florida, west to the eastern

Lesser Siren, Lamar County.

third of Texas, south along the coast into northern Mexico, and north to southwestern Michigan.

NATURAL HISTORY The Lesser Siren prefers warm, shallow, quiet waters such as swamps, sloughs, ponds, lakes, and roadside ditches. Sirens are nocturnal, spending the day burrowed in debris or in mud on the bottom of their aquatic habitats. Because of their aquatic nature, their preference for weed-choked waters, and their nocturnal activity, the presence of these large salamanders usually goes completely unnoticed; however, their population densities can be quite high. Specimens have been collected at a variety of water temperatures, ranging from 8.0°C to 28.0°C (46.5°F to 82.5°F) with a mean of 20.1°C (68°F). Although the external gills are very prominent, the Lesser Siren's lungs are well developed and elongated, containing ridges and folds. The internal surface of the lungs may therefore exceed 3 times the total area of the lung sacs themselves. Approximately 60 percent of gas exchange takes place in the lungs rather than via the prominent external gills. Life in shallow waters presents many challenges. As waters heat up during the warmer months of the year, the dissolved oxygen content of the water becomes greatly depleted. The dual respiratory system of lungs and gills means this salamander is well adapted to survive such conditions. In addition, like most amphibians, it is also capable of respiring across its skin. Another challenge facing the denizens of shallow aquatic environments of Texas is drought. During the hot summer months, many shallow ponds, sloughs, and swamps dry up completely. When the water dries up, the siren burrows into the mud at the bottom, and then forms a parchment-like cocoon that envelops the entire animal except for the mouth. The cocoon is composed of layers of accumulated shed skin or skingland secretions. Such an animal may be effectively entombed for months at a time, protected from desiccation. The siren will remain in its cocoon until the rains return and dry ponds refill. The diet of adults consists of crayfish, worms, mollusks, small fish, and vegetation. It seems that large amounts of aquatic vegetation are engulfed in the course of swallowing animal food. Predators may include water snakes, alligators, various fishes, and wading birds.

REPRODUCTION Reproduction is aquatic. In Texas, Lesser Sirens reproduce in winter or early spring, presumably using external fertilization, with the female depositing eggs in early to late spring. No courtship has been reported, but specimens taken in the breeding season show scars that suggest individuals are aggressive at this time. As many as 200 eggs are laid in burrows or in aquatic vegetation. Eggs are deposited in nests as early as January and on into May. Egg development typically lasts 1.5–2.5 months. Larval development from egg to adult occurs

Lesser Siren, Kaufman County.

Lesser Siren, Anderson County.

in 270 days. Both males and females reach sexual maturity in 2 years.

SUBSPECIES The subspecies in Texas is the Western Lesser Siren, *Siren intermedia nettingi* Goin, 1942.

COMMENTS AND CONSERVATION Lesser Sirens are limited in their western expansion by droughts and the loss of wetland habitats. Flood control programs have probably limited their ability to populate some areas, since flooded bottomlands function as temporary connectors between isolated ponds and allow these animals to disperse between ponds. They are still relatively common in the wetter areas of East Texas and are not a protected species. They are on the TPWD's Black List.

Rio Grande Siren
Siren species "Rio Grande"

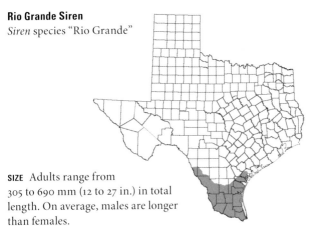

SIZE Adults range from
305 to 690 mm (12 to 27 in.) in total
length. On average, males are longer
than females.

DESCRIPTION Like other siren species, they
are long and slender with 2 small front legs, each with 4 toes.
No hind legs or internal pelvic girdles are present. There are
36–38 costal grooves. The tail is laterally compressed, with a fin
running along the top and bottom edge, and is pointed at the
tip. The dorsal ground color in adults varies from gray to light
gray, brownish gray, or dark gray, with tiny dark spots scattered
over the surface. Juveniles are lighter than the adults, and tend
to darken as they mature.

VOICE Like Lesser Sirens, these sirens may make a noise when
disturbed or when swimming from their burrows toward the
water surface to gulp air. The noise may include yelping, hissing,
croaking, and a sound like that of ducklings.

SIMILAR SPECIES Within its range, the Rio Grande Siren can be
confused only with the Lesser Siren (*Siren intermedia*). The
Lesser Siren has 34–36 costal grooves, while the Rio Grande Siren has 36–38. A large adult with 36 costal grooves would likely
be a Rio Grande Siren, since Rio Grande Sirens are, on average,
much larger than Lesser Sirens.

DISTRIBUTION This species is found from northern Tamaulipas,
Mexico, to the Lower Rio Grande Valley. It has also been recorded up the Rio Grande as far as Webb County, and up the
coast as far as Nueces County.

NATURAL HISTORY The Rio Grande Siren has been found in habitats with permanent and semipermanent water. The habitat con-

Rio Grande Siren, Hidalgo County.

Rio Grande Siren, Hidalgo County.

sists of stationary or slow-moving water, such as that found in ponds, resacas, oxbow lakes, swamps, marshes, ditches, canals, and floodplains. The preferred habitat is usually weed-choked bodies of water with muddy bottoms and deep sediment. This carnivorous salamander feeds on aquatic invertebrates such as crustaceans, insects, insect larvae, worms, and snails, as well as small aquatic vertebrates such as young amphibians, tadpoles, amphibian eggs, and even their own eggs. Herons, snakes, and fish may be predators of this salamander. Swimming away rapidly and biting are common defense mechanisms.

REPRODUCTION Reproduction is aquatic and is thought to be similar to that of *Siren intermedia*. Breeding may occur from late February to March. Around 500 eggs are laid singly or in small groups. Eggs are typically laid in depressions on the bottom of still bodies of water that have mud bottoms and are rich in vegetation. Females may guard their nests.

COMMENTS AND CONSERVATION This "species" has caused a great amount of confusion in the last 30 years. The most recent work grouped the Texas populations with the Greater Siren (*Siren lacertina*), but the general consensus among Texas herpetologists is that these populations are distinct from *S. lacertina*, and work is being done that will place this siren in its own species. Further work needs to be done on the salamander's genetic characterization. This is a species classified as threatened by the State of Texas, which designates it "South Texas Siren (large form), *Siren* sp 1." Development and agricultural activities, such as the draining of lowlands and pesticide use, are the primary causes of the decline of this siren.

ORDER ANURA: FROGS

Frogs and toads, which are collectively called anurans ("anura" means "without a tail"), share a single body form. Almost all lack a tail, and all have 4 legs. The forelegs each have 4 digits; the hind legs are longer and stronger than the forelegs, and each has 5 digits. There is a pronounced hump just behind the middle of the back where the vertebral column joins with the pelvic girdle at the sacrum. All have well-developed eyes with eyelids. The vocal sacs are used only when the male is calling, and each species has a unique call. There are 10 native plus 2 introduced families of anurans found in the United States.

FAMILY BUFONIDAE: TRUE TOADS

Bufonid toads are found almost worldwide, with about 300 species in 34 genera. There are 3 genera in Texas: *Anaxyrus*, *Incilius*, and *Rhinella*. Considered "true toads," bufonids are absent from extremely cold or extremely dry regions and remote oceanic islands. The natural distribution of the family is virtually worldwide, and these toads are found from below sea level in Death Valley in the United States to about 4,800 m (16,000 ft.) in the Andes of South America. They are encountered naturally from the tropics nearly to the Arctic Circle. One species, *Rhinella [Bufo] marina* (Cane Toad), has been introduced into various tropical regions by humans in an effort to control insect pests, but most of these introductions have been miserable failures, as in Australia, where the species has become an ecological pest.

Most true toads have prominent cranial crests and 2 hard black tubercles at the base of each forefoot and hind foot. This family exhibits the typical toad body type: dry, warty skin, short forelimbs and hind limbs used for hopping and walking, and a parotoid gland behind each eye containing toxic substances. The toxic substances in some species can paralyze or kill dogs and other predators and can cause hallucinations in humans. But some predators consume them with no ill effects. For example, hog-nosed snakes (*Heterodon* sp.) and several species of water snakes are adapted to consume toads.

Recognizing a toad is not normally a challenge, but identifying the different species may be a daunting task. Identification is complicated by the tendency of certain species to hybridize with others. Human alteration of habitats has accelerated this condition by bringing previously isolated species closer together. Small toads that have recently transformed from larva are often virtually impossible to identify. In the United States, species may be identified by differences in color, coloration pattern, size, shape of the parotoid glands, prominence and arrangement of the cranial crests, wartiness, and appearance of the foot tubercles.

Breeding generally takes place in the spring or summer, often after rains, and the small-yolked eggs are characteristically laid in long strings on the bottom of shallow, open, semipermanent or permanent bodies of water. They may mate more than once a year and, collectively, during any month. Tadpoles are dark in color and have a short, plump body. Adult males of most species are distinguished by having dark throats. All male toads develop dark nuptial pads on the thumb and inner fingers, which help them cling to the slippery bodies of females during amplexus. Amplexus is pectoral in this family. The vocal sac (species can have 0, 1, or 2 vocal sacs) of most species is round when inflated; in others, it is sausage shaped. In most species, the females grow larger than the males.

American Toad
Anaxyrus [Bufo] americanus,
(Holbrook, 1836)

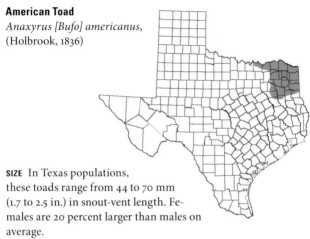

SIZE In Texas populations, these toads range from 44 to 70 mm (1.7 to 2.5 in.) in snout-vent length. Females are 20 percent larger than males on average.

DESCRIPTION These toads have a round, short body and a broad head with kidney-shaped parotoid glands that are usually separated from the cranial crests or connected by short spurs. There are enlarged warts on the tibia. The vocal sac is round when inflated. The background color can range across various shades of brown, gray, or olive; some individuals have a reddish overall appearance. The dorsal side may have spots that are brown or black with yellow, orange, red, or dark brown warts. Each spot contains only 1 or 2 warts. Dark pigment may occur on the chest and forward part of the abdomen. There is usually a light middorsal stripe present. Males have dark-colored throats, and females are lighter overall and have white throats. Pupils are oval and black with a circle of gold around them. There are 4 toes on each front leg and 5 toes connected together by webbing on each hind leg.

VOICE The call is a rather long, high-pitched, sustained musical trill and may last 6–30 seconds, with a trill rate of about 30–40 individual pulses per second.

SIMILAR SPECIES Woodhouse's Toad (*Anaxyrus woodhousii*) has a plain belly, and its warts are more numerous and nearly all the same size. In the American Toad, the cranial crest is separate or else barely touches the parotoid gland with a spur. In the Woodhouse's Toad and Fowler's Toad (*A. fowleri*), the cranial crest comes in direct contact with the parotoid gland.

DISTRIBUTION The American Toad ranges from southeastern and south-central Canada southward into the Carolinas, northern Georgia, Alabama, Mississippi, adjacent areas of Louisiana, and then westward into northeast Texas and eastern Oklahoma and Kansas. There are isolated colonies in southeastern North Dakota, northeastern North Carolina, and Newfoundland.

NATURAL HISTORY These toads have been found in a wide variety of habitats, including gardens, fields, lawns, barnyards, river bottoms, and forest edges. They require shallow bodies of water in which to breed, moist hiding places, and an abundant supply of food items. They shelter from the heat under rocks and logs near preferred habitats, and in cold weather they dig backward into moist soil to hibernate. They are most active when the weather is warm and humid. This crepuscular to nocturnal toad preys on large quantities of beetles, crickets, leaf hoppers, snails, grasshoppers, spiders, moths, slugs, earthworms, and ants. Predators of tadpoles have been reported to be predaceous diving beetles, newts, dragonfly naiads, giant water bugs, crayfish, and sandpipers. Recorded predators of adults include hog-nosed snakes, water snakes, ducks, crows, screech owls, raccoons, skunks, and opossums.

REPRODUCTION Breeding occurs from February to July in shallow temporary ponds, ditches, or shallow, slow streams. At breeding sites are a number of males that do not call, but attempt to in-

American Toad, Red River County.

American Toad, Wood County.

tercept females as they arrive. During reproduction, males have enlarged horny pads on the inside of each forelimb and the presence of a dark gray to black throat. From 2,000 to 20,000 eggs are laid in water 50–100 mm (2.0–3.9 in.) deep, in large clutches, with 2 long strings attached to vegetation; eggs hatch in 3–12 days. Metamorphosis occurs between June and August and lasts 40–70 days.

SUBSPECIES The subspecies found in Texas is the Dwarf American Toad, *Anaxyrus americanus charlesmithi* (Bragg, 1954).

COMMENTS AND CONSERVATION This toad is on the TPWD's Black List. In the northeastern United States, this toad does not appear to be sensitive to alterations in habitat, but may be subject to fluctuations in population size.

Great Plains Toad
Anaxyrus [Bufo] cognatus,
(Say, 1823)

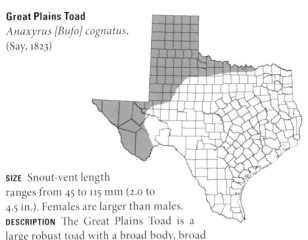

SIZE Snout-vent length
ranges from 45 to 115 mm (2.0 to
4.5 in.). Females are larger than males.
DESCRIPTION The Great Plains Toad is a
large robust toad with a broad body, broad
head, and short snout. A boss between the eyes is formed by
prominent V-shaped cranial crests that converge diagonally
from the back of its head. It has elongated oval parotoid glands
behind the cranial crests. The hind legs are about the same
length as the body. A sharp-edged tubercle is located on each
of the hind legs. The dorsal ground color may be yellowish, light

Great Plains Toad, Moore County.

brown, greenish, olive, or gray, with dusky, dark gray, olive, or green blotches bordered by light pigment in symmetrical pairs down the back. A narrow light-colored middorsal stripe may be present. The venter is whitish, usually unspotted. When fully inflated, the sausage-shaped vocal sac may be up to one-third the size of the toad.

VOICE The call is a harsh, explosive, piercing, shrill metallic trill lasting 5–60 seconds at a rate of 13–20 individual pulses per second.

SIMILAR SPECIES Although the Texas Toad (*Anaxyrus speciosus*) is similar in appearance and may have pairs of spots on its back, the spots are smaller and less well defined than in the Great Plains Toad. In addition, the cranial crests and boss on the snout of the Texas Toad are less prominent, and the voice is a series of short trills rather than a prolonged trill.

DISTRIBUTION The range coincides with the Great Plains. It is found from extreme southern Canada southward through the United States to as far south as San Luis Potosí, Mexico, and from west Texas westward into extreme southeastern California and southern Nevada. Vertically, it ranges from sea level to 2,440 m (8,000 ft.). There is a spotty distribution in the desert part of its range.

NATURAL HISTORY The habitat of the Great Plains Toad includes short and tall grasslands, open floodplains, mesquite woodlands, sagebrush plains, cultivated areas, areas near man-made construction, and prairies or deserts, but rarely in upland wood-

Great Plains Toad, Jeff Davis County.

Great Plains Toad, Presidio County.

lands. They breed after heavy summer rains in shallow temporary pools or the quiet water of streams, marshes, irrigation ditches, and flooded fields, where they may gather in large numbers. These toads can tolerate drier conditions and man-made disturbed areas better than most other species. They are accomplished burrowers. This toad is usually crepuscular or nocturnal, but may be active on cloudy or rainy days. This opportunistic feeder may feed on cutworms, beetles, ants, moths, flies, crickets, termites, spiders, centipedes, and mites. Tadpoles feed on algae and decomposing invertebrate remains. Defensive posturing includes inflating the body, closing the eyes, and lowering the head to the ground to expose its toxic parotoid glands and warts. This posture also makes the toad look like a rocklike or clod-like object. Tadpoles are taken by birds, insect larvae, and other tadpoles. Adults are taken by a large variety of mammals, birds, and snakes.

REPRODUCTION The Great Plains Toad generally breeds from March to September in shallow temporary pools formed after heavy rains. Eggs are laid in long single or double strings of 1,400–45,000 eggs attached to vegetation and debris. Eggs hatch in 2–7 days.

COMMENTS AND CONSERVATION The Great Plains Toad is one of the most common amphibians on the High Plains of Texas. Their fossorial nature makes monitoring their populations difficult. They are on the TPWD's White List. The Great Plains Toad is considered an important support to agriculture because of its consumption of large numbers of cutworms in some areas.

Green Toad
Anaxyrus [Bufo] debilis,
(Girard, 1854)

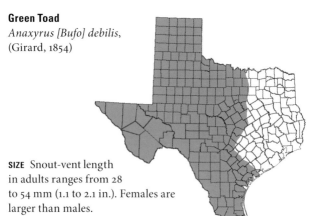

SIZE Snout-vent length in adults ranges from 28 to 54 mm (1.1 to 2.1 in.). Females are larger than males.

DESCRIPTION *Anaxyrus debilis* is a small flat green, pale green, or yellow-green toad with small black spots and bars on its back. The black markings on the back may be united to form a weblike, or reticulated, pattern. Both males and females have a white ventral surface; the male has a dark throat, while the females have a yellowish throat. The parotoids are large, elongated, and widely separated, extending downward to the jaw. Cranial crests are absent or very faint. The vocal sac is round.

VOICE The call is similar to that of Narrow-mouthed Toads, but even louder and shriller. It consists of a 2–10 second buzz that sounds like an electric buzzer.

SIMILAR SPECIES Some small Texas Toads (*Anaxyrus speciosus*) are greenish in color, but never as green as Green Toads. And Texas Toads do not have the black spots or bars on the back.

DISTRIBUTION Green Toads range from southeastern Colorado and southwestern Kansas southward into northern Zacatecas, Tamaulipas, and San Luis Potosí, Mexico, and from East Texas to southeast Arizona. These toads have been found at elevations as high as 1,830 m (6,000 ft.).

NATURAL HISTORY This toad has been found in arid and semiarid grasslands, mesquite–short grass prairies, playa bottom grasslands, mesquite and creosote brush, mesquite savannas, and the valleys of small creeks. During the day, they take refuge under rocks or in rodent burrows near temporary water sources, including stock tanks, rain pools, roadside ditches, or shallow pools in intermittent streams. They are seldom seen aboveground except during and after periods of heavy summer rains. Tadpoles feed on a variety of algae and dead organic matter on or near the bottom of temporary pools. The adults take a variety of prey items: ants, small butterflies and moths, small beetles, and small grasshoppers. Known predators include American Bullfrogs, Tiger Salamanders, and several species of garter snakes. Tadpoles are probably taken by a wide variety of predaceous aquatic insects such as backswimmers, giant water bugs, predaceous diving beetles, and water scavenger beetles. Adults are probably taken by snakes, skunks, ravens, and raccoons. The parotoid gland produces toxic steroids that may render the toads unpalatable to some predators.

REPRODUCTION Breeding takes place from March until September and peaks during May in the eastern part of the range, and peaks during July in the western part of the range. Breeding occurs in temporary streams and pools that form during spring or summer rains. If rains do not occur or conditions are not favorable, breeding may not occur. Because of the temporary nature of the breeding pools that this toad uses in arid and semiarid habitats, tadpoles transform quickly. The nonadhesive eggs are laid in small clumps or strings, or are singly attached

Green Toad, Irion County.

Green Toad, Culberson County.

to grass and weed stems. Hatching occurs in 1–7 days. Development ranges from 8 to 25 days, depending on various environmental factors.

SUBSPECIES Two subspecies of Green Toad are recognized in Texas, the Eastern Green Toad (*Anaxyrus debilis debilis* [Girard, 1854]) and the Western Green Toad (*A. d. insidior* [Girard, 1854]).

COMMENTS AND CONSERVATION These toads are on the TPWD's White List. Green Toads tend to be localized but common in available habitats. Because of habitat alteration, these toads have declined in some areas.

Fowler's Toad

Anaxyrus [Bufo] fowleri,
(Hinckley, 1882)

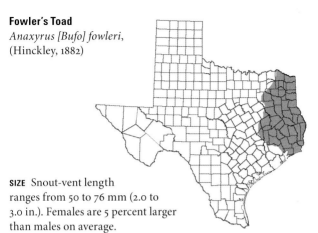

SIZE Snout-vent length
ranges from 50 to 76 mm (2.0 to
3.0 in.). Females are 5 percent larger
than males on average.

DESCRIPTION The color of this toad varies
widely in specimens from different locales. General dorsal coloration may be uniformly yellowish brown, brown, greenish
brown, or nearly black, with many having an overlaying of reddish wash. A Texas Fowler's Toad may have the following characteristics: a dark pectoral region that breaks apart into dark
spots on the belly, a narrow light middorsal stripe, vague lateral
light stripes (occasionally), cranial crests that touch the parotoid
glands behind the eye, 3 or more warts in each large spot on the
dorsal surface, no enlarged warts on the tibia, and a supratympanic ridge. Fowler's Toads from some areas may have all these
characteristics, whereas others may only have one.

VOICE The call of this toad varies widely. Coming at intervals of
1–5 seconds, it is similar to the muffled bleat of a young sheep.
The call sounds something like *waaaaaaaah.*

SIMILAR SPECIES The belly of the Fowler's Toad is highly spotted,
the dorsum is darker, and the call is somewhat different from
those characteristics of Woodhouse's Toad (*Anaxyrus woodhousii*). Fowler's Toad has a supratympanic ridge not found in
other toad species in Texas.

DISTRIBUTION The range includes the eastern United States from
southern New England southward to the Gulf Coast, westward
to Michigan, Illinois, Missouri, and south to southeastern Oklahoma and eastern Texas. It is absent from eastern South Carolina, southern Georgia, and most of Florida.

NATURAL HISTORY The nocturnal Fowler's Toad prefers sandy areas near marshes, irrigation ditches, temporary rain pools, deciduous woodlands, roadside ditches, fields, and pastures; it may also be seen in backyard gardens. These toads breed in shallow water in flooded areas, permanent ponds, low-running streams, lakeshores, and temporary pools, or along the shallow areas of rivers. They avoid desiccation during hot and dry periods by burrowing into the ground or finding mammal burrows. Their prey primarily includes beetles and ants. They usually approach their prey by walking instead of hopping. This toad is often seen under lights, feeding on insects. Predators include hognosed snakes, some birds, American Bullfrogs, and raccoons. The toad's toxic or noxious skin secretions may help in its defense against most predators.

REPRODUCTION Breeding may take place from March to August, depending on moisture levels. Most often this occurs in March and April after early spring rains. From 2,000 to 4,000 eggs are laid in long twin strings attached to vegetation in shallow water. Large aggregations of tadpoles are sometimes observed.

COMMENTS AND CONSERVATION This toad is recognized as *Anaxyrus woodhousii* by the TPWD and is on its White List. In some areas, this toad has declined dramatically. The exact reason is not known, but most likely is the alteration of habitats from urbanization and the drainage of breeding pools. Controversy has

Fowler's Toad, San Augustine County.

Fowler's Toad, Tyler County.

been associated with this taxon for many years. Many authorities recognize the Texas population as hybrid in origin between *A. woodhousii* and *A. fowleri*. Others consider this organism to a subspecies (*A. w. velatus*). Some have considered it to be a valid species, the East Texas Toad (*A. velatus*). We recognize these East Texas toads as belonging to the more widespread eastern species, Fowler's Toad (*A. fowleri*), because of the overall similarities between Texas populations and those from the southeastern United States.

Houston Toad

Anaxyrus [Bufo] houstonensis,
(Sanders, 1953)

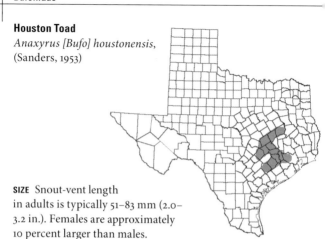

SIZE Snout-vent length in adults is typically 51–83 mm (2.0–3.2 in.). Females are approximately 10 percent larger than males.

DESCRIPTION Background color can be light brown, gray, purplish gray, or reddish, with darker patches of green, dark green, brown, or black. These darker patches on the dorsal surface can be in a mottled pattern arranged in a vaguely herringbone or zigzag fashion. A middorsal light stripe is usually present. The ventral surface is pale, usually with numerous small dark spots, especially between the front limbs. There is always at least one brown spot on the pectoral region. Males and females are similar, except males have a dark throat. The vocal sac on males is round when inflated. The ridges of the cranial crest are thickened, especially those running between the eyes; spurs on the outside edges of the ridges may come in contact with the elongated parotoid glands.

VOICE The 4–11 second call is a high-pitched musical trill with a rate of 32 individual pulses per second. The call is higher pitched than that of the American Toad (*Anaxyrus americanus*). After amplexus, the males issue a release call consisting of a short, barely audible vibration and an even shorter chirp.

SIMILAR SPECIES In Woodhouse's Toad (*Anaxyrus woodhousii*) and Fowler's Toad (*A. fowleri*), the parotoid gland touches the cranial ridge behind the eye. The Gulf Coast Toad (*Incilius nebulifer*) has a pronounced dark stripe along the side of the body and a deep valley between the eyes. In the Texas Toad (*A. speciosus*), both tubercles beneath the hind foot have sharp cutting edges.

DISTRIBUTION This toad is found in isolated populations in south-central and southeastern Texas, ranging from Harris County westward into Bastrop County. The southeastern population has apparently been extirpated, since no recent sightings have been reported from this region.

NATURAL HISTORY These toads are restricted to deep sandy soils with subsurface moisture. Loblolly pine forests, mixed deciduous forests, post oak savannas, and coastal prairies are preferred areas. These nocturnal toads spend the day burrowed in sand and are never found far from their breeding areas—nonflowing rain pools, flooded fields, roadside ditches, and ponds. Tadpoles are nocturnal, living at the bottom of their pools in the daytime and moving to the surface at night to feed. Prey items of the adult include beetles, flies, lacewings, moths, ants, and other toads or frogs. They use both active-search and sit-and-wait techniques to obtain their prey. Tadpoles feed on pollen (usually from nearby pine trees), the jelly envelopes of other recently hatched Houston Toads, and algae on floating leaves. Water snakes, hog-nosed snakes, herons, egrets, raccoons, skunks, and coyotes are the primary predators of these toads. Fire ants have been reported to attack and consume young terrestrial toads.

REPRODUCTION Breeding activity occurs from the end of January to the end of June, with the peak usually in early March, when rains create shallow pools. During some wet springs, breeding

Houston Toad, Bastrop County.

Houston Toad, Leon County.

may be repeated, occurring first in late February or early March and then again in late May or early June. Males call from or near grass-rimmed pools that persist for at least 60 days. Typically, 500–6,000 eggs are deposited in strings during the early-morning hours. Eggs hatch in about 2–7 days. Metamorphosis begins in 53–65 days.

COMMENTS AND CONSERVATION The Houston Toad is threatened by extinction because of modification of its habitat by urbanization, agricultural practices, prolonged droughts during the 1950s, recreational overdevelopment, road mortality, deforestation, and lignite-coal mining operations. The Houston Toad is on the Texas and federal endangered species lists. Conservation activities have included land acquisitions by the state, reintroducing toads into areas where once present, and continued research on the toad's life history. Recently conservation efforts, namely, the head starting of toadlets in captivity, are having some success.

Red-spotted Toad
Anaxyrus [Bufo] punctatus,
(Baird and Girard, 1852)

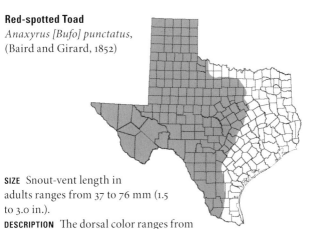

SIZE Snout-vent length in
adults ranges from 37 to 76 mm (1.5
to 3.0 in.).

DESCRIPTION The dorsal color ranges from
light gray, olive, or light brown to medium
brown, reddish brown, and pale olive, with buff, reddish, or orange warts, which are at times set in small dark blotches. Many specimens from the limestone country of the Edwards Plateau are pale gray and virtually unmarked. The ventral surface is buff or white, with or without spotting; males have a dusky throat. Young individuals usually have numerous red or orange-tipped

Red-spotted Toad, Culberson County.

Red-spotted Toad, Edwards County.

warts and are dark below. The undersides of the feet are yellow. This toad can be distinguished by its flattened head and body, small round parotoid glands no larger than the eye, pointed snout, and weak or absent cranial crests. The vocal sac is round.

VOICE The call is a musical trill, a high-pitched monotone lasting about 4–10 seconds. The interval between calls is variable and may be longer or shorter than the call itself. These toads usually call from rocks or gravel near the water's edge.

SIMILAR SPECIES All other Texas toads have elongated parotoid glands and well-defined cranial crests, or both; this is the only toad in Texas with round parotoid glands.

DISTRIBUTION Red-spotted Toads range from southern Nevada, southern Utah, Arizona, southeastern California, western and southern Oklahoma, western Texas, and southwestern Kansas southward into Mexico, including Baja California.

NATURAL HISTORY The habitat of the Red-spotted Toad includes rough, rocky regions of desert streams and oases, open grasslands and scrublands, mammal burrows, oak woodlands, and rocky canyons and arroyos. This toad is common in areas of limestone, where it is an agile climber. During the day, Red-spotted Toads find shelter in crevices and under flat rocks. In the desert Southwest, these toads rarely venture far from their breeding habitats: springs, seepages, persistent pools along streams, cattle tanks, or rain-formed pools in upland areas. They have been found from below sea level to around 1,980 m (6,500 ft.). Red-spotted Toads are mostly nocturnal except during breeding season, when they may be active anytime, especially during rainy periods. The toad's physiology allows it to absorb nearly 70 percent of its water through a specialized patch of skin on the rear end when exposed to a small and transient water source. A wide variety of invertebrates are included in their diet. Predators have been recorded as snakes, birds, and small mammals.

REPRODUCTION The Red-spotted Toad breeds from March to September. In the eastern part of their range, they breed from March to June in association with spring rains. In the desert Southwest, they breed from June to September in association with summer monsoonal rains. An average of 1,500 sticky, pigmented eggs are laid singly, in short strings, or as a loose flat cluster, on the bottom of small, shallow, often rocky pools. Eggs hatch in about 3 days. Tadpoles may form large aggregations and usually transform after 40–60 days.

COMMENTS AND CONSERVATION This species is on the TPWD's White List. It is relatively common in most areas of its range.

Texas Toad

Anaxyrus [Bufo] speciosus,
(Girard, 1854)

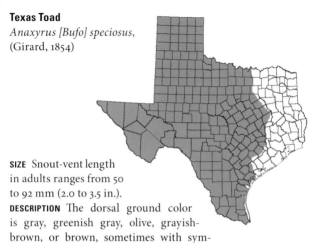

SIZE Snout-vent length in adults ranges from 50 to 92 mm (2.0 to 3.5 in.).

DESCRIPTION The dorsal ground color is gray, greenish gray, olive, grayish-brown, or brown, sometimes with symmetrically arranged dark blotches on the back. The back may be marked with yellowish-green or brown spots and pink, orange, or greenish warts. The ventral surface is white to cream, unmarked or with dark spots. Young Texas Toads are gray-brown above, blotched with green, and flecked with black; warts are tipped with red. Males have an olive-colored skin, and the olive-colored deflated vocal sac of males is covered with a pale skin fold. The inflated vocal sac is large and sausage shaped and equals about one-third of the bulk of the toad. The cranial crests are weak or absent. The inner tubercles on the hind foot are sickle shaped and are usually black with sharp edges. Parotoid glands are oval and widely separated.

VOICE The Texas Toad's call is a loud, continuous shrill trill, each call lasting about 0.5–7 seconds at intervals of about 1 second and with a trill rate of 39–57 individual pulses per second. After the toad calls for several minutes, there may be a distinct drop in pitch, followed by a return to the original pitch.

SIMILAR SPECIES The Texas Toad is a close relative of the Great Plains Toad (*Anaxyrus cognatus*) but can be differentiated by its round, plump, rather plain-colored, and uniformly warty nonstriped dorsal surface. The Red-spotted Toad (*A. punctatus*) from the Edwards Plateau may be the same color as Texas Toads, but they have parotoid glands that are smaller and round in shape. Cane Toads (*Rhinella marina*) have parotoid glands that are very large and extend far down onto the side of the body. Woodhouse's Toads (*A. woodhousii*) have prominent cranial crests.

DISTRIBUTION This toad is primarily a Texas species found in the western two-thirds of the state. It ranges into western Oklahoma and southeastern New Mexico, and southward into Chihuahua, Coahuila, and central Tamaulipas, Mexico.

NATURAL HISTORY The habitat of this nocturnal toad may include breeding sites such as permanent streams, irrigation ditches, watering tanks, and ephemeral pools, as well as nonbreeding sites such as mesquite woodlands, prairies and farmlands, grasslands, cultivated areas, and mesquite-savanna associations. It prefers sandy soils, in which it is a skilled burrower. Vertically, it is found from sea level to more than 1,706 m (5,600 ft.). Even though these toads have adapted to dry conditions, it is crucial that they find refuge in sites with high humidity or sites with good water-holding capacity in order to prevent desiccation during long dry periods. The Texas Toad is an opportunistic feeder that takes a wide array of terrestrial and flying arthropods, such as true bugs, various beetles, and ants. Grackles, beetle larvae, and mud turtles feed on tadpoles; garter snakes prey on both adults and tadpoles. When threatened, if it is unable to burrow,

Texas Toad, Brewster County.

Texas Toad, Dawson County.

it will flatten itself against the ground to avoid detection. To deter predators, it secretes toxic substances from both the skin and the parotoid glands.

REPRODUCTION Following heavy spring and summer rains from April through September, the Texas Toad migrates to breeding sites. Multiple females may deposit their eggs at the base of the same clump of vegetation. Hatching occurs in 2 days. Metamorphosis takes 18–60 days.

COMMENTS AND CONSERVATION The Texas Toad is on the TPWD's White List. Throughout the majority of its range in Texas, its population appears to be stable. However, in the Lower Rio Grande Valley, where its numbers have declined in the last 40 years, it seems to be affected by heavy pesticide and herbicide use. The Texas Toad may hybridize with Woodhouse's Toad (*Anaxyrus woodhousii*) and the Great Plains Toad (*A. cognatus*).

Woodhouse's Toad

Anaxyrus [Bufo] woodhousii,
(Girard, 1854)

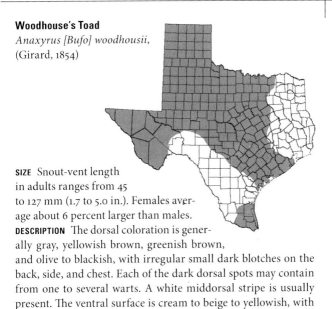

SIZE Snout-vent length
in adults ranges from 45
to 127 mm (1.7 to 5.0 in.). Females average about 6 percent larger than males.

DESCRIPTION The dorsal coloration is generally gray, yellowish brown, greenish brown,
and olive to blackish, with irregular small dark blotches on the
back, side, and chest. Each of the dark dorsal spots may contain
from one to several warts. A white middorsal stripe is usually
present. The ventral surface is cream to beige to yellowish, with
or without dark flecks. There may be a network of yellow and
black markings on the rear of the thighs. The male has a dark

Woodhouse's Toad, Moore County.

throat that is set off by a pale yellow border on the lower jaw. The inflated vocal sac is round. A prominent narrow cranial crest, usually touching the elongated, divergent parotoid gland, is located just behind the eyes. Some have a boss between the eyes.

VOICE The sheep-like bleating call is described as an explosive, nasal *waaaaaah* that lasts 1-4 seconds, often suddenly dropping in pitch at the end. Males call while sitting in shallow pools of still water and on land near the water.

SIMILAR SPECIES Some young Woodhouse's Toads may have reddish-tipped warts that are similar in appearance to those of the Red-spotted Toad (*Anaxyrus punctatus*), but Woodhouse's Toad generally has elongated parotoid glands and a middorsal stripe. The Red-spotted Toad also lacks a boss between the eyes. The American Toad (*A. americanus*) has numerous dark markings on the chest, and the cranial crest does not come in contact with the parotoid gland except via a spur from the outer edge.

DISTRIBUTION This toad ranges from eastern Oregon and California, across to eastern Montana and the Dakotas, south through Utah, Colorado, Nebraska, Kansas, Oklahoma, Texas, New Mexico, and Arizona, and on into Durango, Mexico. The toads in South Texas may be isolated from the main population.

NATURAL HISTORY The primary habitat of this toad includes grasslands, sagebrush flats, woods, desert streams, valleys, floodplains, farms, ditches, irrigated fields, cultivated areas, marshes, river bottoms, mountain canyons, and even suburban backyards, as long as there is sufficient moisture available. There is

Woodhouse's Toad, Oldham County.

Woodhouse's Toad, Oldham County.

a preference for sandy areas. In the western part of their range, they prefer riparian corridors at lower elevations, and moist meadows, ponds, lakes, and reservoirs at higher elevations. Normally nocturnal, they may be seen feeding under streetlights or yard lights in suburban areas. During the day, they hide under rocks, debris, or vegetation, or in burrows. Prey includes a large variety of invertebrates, especially beetles. Bull snakes, rat snakes, American Bullfrogs, hawks, skunks, and roadrunners have been observed feeding on these toads. Presumably, their skin secretions afford some protection against some predators, both as adults and as tadpoles.

REPRODUCTION This toad may breed from February to September, depending on the part of the country it is found in. In western grasslands and desert areas, breeding occurs usually after rains. In the eastern part of their range, breeding usually occurs during spring rains. In riparian zones in the West, they may breed anytime during rainless spring nights. More than 25,000 pigmented eggs are laid in long gelatinous strings in 1 or 2 rows, intertwined about vegetation or debris, in almost any type of shallow pool or stream.

SUBSPECIES The subspecies found in Texas are the Rocky Mountain Woodhouse's Toad (*Anaxyrus woodhousii woodhousii* [Girard, 1854]) and the Southwestern Woodhouse's Toad (*A. w. australis* [Shannon and Lowe, 1955]). Woodhouse's Toads may hybridize with the Gulf Coast Toad (*Incilius nebulifer*), the Texas Toad (*A. speciosus*), and the Houston Toad (*A. houstonensis*).

COMMENTS AND CONSERVATION The isolated population of these toads in South Texas probably has been extirpated (except on Padre Island) through hybridization events and pesticide applications to farmland. Until recently, these toads did not seem to be having any population declines in the state, except in South Texas and along the Rio Grande south of El Paso, but now they may be declining along the eastern edge of their range. The western subspecies (*A. woodhousii australis*) has not been encountered in Texas for many years. This species is on the TPWD's White List.

Gulf Coast Toad
Incilius [Bufo] nebulifer,
(Girard, 1854)

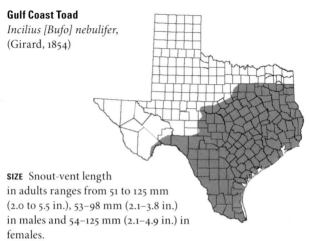

SIZE Snout-vent length
in adults ranges from 51 to 125 mm
(2.0 to 5.5 in.), 53–98 mm (2.1–3.8 in.)
in males and 54–125 mm (2.1–4.9 in.) in
females.

DESCRIPTION This is a broad-headed, flat toad with prominent cranial crests that form a depression on top of the skull. The triangular parotoid glands are connected to the cranial crests behind the eyes. The vocal sac becomes large and rounded during calling. There are prominent broad dark lateral stripes bordered above by a light stripe, and there is also a light middorsal

Gulf Coast Toad, Williamson County.

stripe. Background color varies from almost black to brown, yellow brown, or gray, with whitish or orange spots. Males have a yellowish-green throat. Above the pale lip area, a distinct narrow dark line runs the length of the upper lip. The ridges of the crests may be dark. The normal resting posture tends to be squattier than that seen in other species.

VOICE The call is a short, flat trill, lasting 2–6 seconds, and is repeated several times at intervals of about 1–4 seconds.

SIMILAR SPECIES No other toad in Texas has dark prominent lateral lines or deep prominent cranial crests.

DISTRIBUTION The species ranges from north-central Texas and the western edge of the Edwards Plateau of Texas, southern Arkansas and eastern Louisiana, southward into central Veracruz, Mexico.

NATURAL HISTORY These ubiquitous, crepuscular to nocturnal toads are at home in a variety of moist habitats, including lawns and under streetlights, as well as in railroad ditches, roadside pools, irrigation ditches, coastal prairies, dump sites, storm sewers, and cave mouths. They also reside on barrier islands along the Gulf of Mexico. Tadpoles feed on algae and on animal material found on substrates. Prey items of adults may include isopods, beetles, lizards, other toads, scorpions, and other available

Gulf Coast Toad, Brazos County.

small animals. Adults may be preyed on by Tiger Salamanders, Western Ribbon Snakes, water snakes, Cat-eyed Snakes, and Indigo Snakes. Diving beetles are the primary predator of tadpoles. As with other toads, Gulf Coast Toads are distasteful to a wide range of potential predators, presumably because of their toxic parotoid gland secretions.

REPRODUCTION This toad primarily breeds in the springtime, but may breed any time from February to September after heavy rains. They will reproduce in a wide variety of habitats, from temporary pools to permanent bodies of water. Gulf Coast Toads generally deposit eggs in strings of jelly, usually in double rows. A single clutch may contain as many as 20,000 eggs, which hatch in 1–2 days. Tadpoles transform after 20–30 days.

COMMENTS AND CONSERVATION The Gulf Coast Toad is on the TPWD's White List, and it is one of the most abundant amphibians within its range. Their adaptability and ability to tolerate habitat alterations makes this species able to thrive in an urban environment. In Central Texas, Gulf Coast Toads are now found more abundantly than Texas Toads and Woodhouse's Toads, which were formerly more abundant in the region.

Cane Toad
Rhinella [Bufo] marina,
(Linnaeus, 1758)

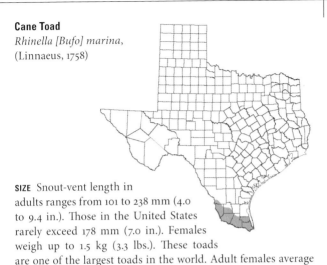

SIZE Snout-vent length in adults ranges from 101 to 238 mm (4.0 to 9.4 in.). Those in the United States rarely exceed 178 mm (7.0 in.). Females weigh up to 1.5 kg (3.3 lbs.). These toads are one of the largest toads in the world. Adult females average 13 mm (0.5 in.) larger than males.

DESCRIPTION The body shape is round and flattened. Coloration varies from dark brown to light brown or tan, with yellow, red, or even olive-green tinting. A light middorsal stripe may be present, and there may be a faint to dark pattern of slightly darker blotches. The granular ventral surface tends to be whit-

Cane Toad, Hidalgo County.

ish yellow with dark brown specks, or may be mottled. This very large brown toad is characterized by large, deeply pitted parotoid glands that extend far down the sides of the body. There are prominent cranial crests, which give an appearance of horned ridges. On either side of the midline there may be a row of large, fleshy warts running the length of the body. These warts are more prominent on males. The inflated vocal sac is round.

VOICE The call, which can be heard great distances, is a slow, rhythmic, low-pitched trill. The sound is reminiscent of the exhaust noise of a distant tractor or outboard motor.

SIMILAR SPECIES Because of its huge parotoid glands, it is not likely to be confused with other frogs or toads where it is found. Young toads may be confused with Gulf Coast Toads (*Incilius nebulifer*). Gulf Coast Toads have much smaller parotoid glands as well as dorsal and lateral stripes.

DISTRIBUTION Its natural range barely enters the United States— only in extreme southern Texas. It is found in western Mexico to southern Sonora, and southward to and through Central and South America into the central Amazon Basin. It has been introduced into many tropical parts of the world, such as Jamaica, the Philippines, Puerto Rico, Fiji, New Guinea, and Australia, as well as Florida and Hawaii.

NATURAL HISTORY The habitat of this nocturnal, terrestrial toad includes pools, arroyos, gardens, residential yards, water tanks, disturbed areas, riversides, and coasts. Along the coast, they may be found in association with both fresh and brackish water, including mangrove swamps. Juveniles are mostly diurnal and

Cane Toad, Starr County.

Cane Toad, Hidalgo County.

rarely wander far from a water source. During the day, these toads seek shelter under stones, logs, or boards or in burrows that they dig in soft earth. Their diet is limited only by the gape of their jaws. These toads are nonspecific feeders and will consume almost any organism smaller than they are. Food items include beetles, cockroaches, crabs, spiders, centipedes, millipedes, scorpions, caterpillars, snails, slugs, frogs, toads, snakes, birds, mammals, and even dog food. They have been reported to eat vegetable matter occasionally. The tadpoles are almost exclusively algae feeders. The Cane Toad is resistant to predation because of the highly toxic, foul-tasting, sticky, milky substance secreted from its skin and glands. The secretion has been known to cause skin and eye irritation in humans and can be lethal to an animal that bites the toad. Natural predators include toad-eating snakes, small mammals, and birds. There are fewer predators in areas were the toad has been introduced. Head-down defensive posturing has been reported. The toad flattens the body and tilts it toward the aggressor by straightening the 2 legs on the far side. Cane Toads have been reported to live for more than 15 years in captivity.

REPRODUCTION Reproduction is strictly aquatic and takes place whenever rainfall and temperature are optimal. The Cane Toad will breed in most types of brackish or fresh water, including temporary pools, ponds, ditches, canals, and streams. It lays from 25,000 to more than 30,000 black eggs in long, jellylike strings on rocks, debris, and vegetation in standing water. The egg-strings can measure up to 18 m (59 ft.) in length. Eggs take 3–7 days to hatch. After hatching, the small black tadpoles aggregate in dense numbers. They metamorphose in 12–60 days.

COMMENTS AND CONSERVATION In Texas, the Cane Toad is on the TPWD's Black List. While some other toad species are having population declines, the Cane Toad is increasing the size of its range.

FAMILY HYLIDAE: TREE FROGS

Hylidae is a large family composed of more than 901 species throughout the world in 46 genera. They are found on all continents except Antarctica. They are most abundant in the tropics of the New World. Five genera in this family occur in Texas: *Hyla* (tree frogs), *Acris* (cricket frogs), *Pseudacris* (chorus frogs), *Smilisca* (Mexican tree frogs), and the recent introduction *Osteopilus* (*O. septentrionalis*, the Cuban Treefrog). Tree frogs are usually arboreal, with usually small, slim waists, long legs, and well-developed toe pads that are set off from the rest of the toe by a small extra phalanx. Cricket frogs and chorus frogs have much smaller toe pads and spend most of their lives near or on the ground among thick vegetation. Except during the breeding season, chorus frogs are very secretive and venture forth only after or during rainy periods. Depending on conditions of light, moisture, temperature, stress, or activity, these frogs may change from a pale gray, green, or brown to either a more vivid solid color or to a bold, distinct pattern. The young of several species may exhibit a plain, bright green pattern for long periods of time, making it difficult to tell them apart. Some species of tree frogs have distinctive markings or colorations on the concealed surfaces of their hind legs, making capture and close examination critical for identification. Male tree frogs have a vocal sac that looks like a round balloon under the throat when inflated. The only exceptions among species in the United States are the Mexican Treefrogs and Cuban Treefrogs, in which a single sac inflates more on each side than in the middle, thus producing the suggestion of a double sac.

Northern Cricket Frog
Acris crepitans, Baird, 1854

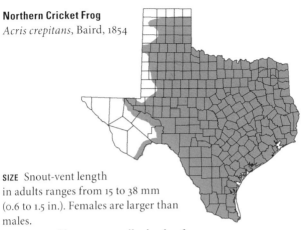

SIZE Snout-vent length
in adults ranges from 15 to 38 mm
(0.6 to 1.5 in.). Females are larger than
males.

DESCRIPTION These are small, slender frogs
with a slim waist and slender, extensively webbed toes (without
toe pads). There is a triangular mark on the head (the marking
occasionally may be absent), with the apex pointing backward.
The dorsal color is gray, light brown, green, reddish, or reddish
green, and usually there is a similarly colored middorsal stripe.
The back has dark markings, and there are dark bands on the
legs. Two excellent distinguishing marks are a white bar that ex-

Northern Cricket Frog, Mason County.

Northern Cricket Frog, Edwards County.

tends from the eye to the base of the foreleg, and a dark stripe on the rear of the thigh. Males have a dusky throat, suffused with yellow, with more spotting than that found on females. The throat of the male darkens in breeding season. The vocal sac is single and round. The skin is rough or bumpy.

VOICE The call has been described as a metallic *gick-gick-gick*, resembling the sound made by striking 2 river rocks together. There is about 1 call per second for 20–30 or more beats, tapering in frequency at the beginning and end of a call.

SIMILAR SPECIES The Upland Chorus Frog (*Pseudacris feriarum*) has a whitish stripe along the upper lip and a lengthwise brownish stripe on the sides and back, and the toes are slightly webbed. Young *Lithobates* frogs, which are often mistaken for Cricket Frogs, may be differentiated by the latter's thigh stripe, triangular mark on the head (which is sometimes absent), and white facial bar. The Northern Spring Peeper (*P. crucifer*) has smooth skin and an X-shaped marking on the back.

DISTRIBUTION The Northern Cricket Frog has a wide distribution over the eastern and central United States, ranging into the short grass plains of eastern Colorado and New Mexico. It is found from Michigan southward into northeastern Mexico, and from Long Island, New York, westward into eastern Colorado and southeastern New Mexico. It occurs throughout most of Texas except for the extreme northern Panhandle and the western part of the Trans-Pecos. Vertically, it ranges from sea level to around 1,220 m (4,000 ft.).

NATURAL HISTORY *Acris crepitans* is a diurnal and nocturnal non-climbing member of the tree frog family. It prefers ponds with ample vegetation in the water and on land, and full sun during most of the day. It may be encountered also in and around slow-moving streams with sunny banks. This species will also live in brackish water habitats in coastal marshes within a mile of the Gulf of Mexico. This frog spends most of its time on the ground, hopping among the grass and other vegetation. The Northern Cricket Frog is active all year, except in midwinter in the north, and is often found in groups. Individuals scatter when frightened, making selection of a single frog to pursue difficult. If disturbed, it will leap quickly out of reach or skitter over the water's surface. It can jump more than 1 m (3 ft.), and often uses a zig-zag pattern. Predators include snakes, American Bullfrogs, turtles, fish, herons, other birds, and minks. The diet of adults consists of insects and invertebrates, including mosquitoes, spiders, annelids, mollusks, crustaceans, and plant matter. Tadpoles feed on various types of algae.

REPRODUCTION Since the life span averages 4 months, within 6 months the Northern Cricket Frog population can be completely replaced by the next generation. Some have been recorded to live up to 3 years. The cricket frog breeds from February to August in the North and from February to July in the

Northern Cricket Frog, San Jacinto County.

Northern Cricket Frog, Burnet County.

Northern Cricket Frog, Leon County.

South. Choruses can be heard as late as October in its southern range. They search out semipermanent and permanent wetlands and river backwaters for breeding. Eggs are laid singly or in small clusters, and are attached to leaves, twigs, or grass stems, or else are laid on the bottom of springs, ponds, and streams in shallow, quiet water. A single female may lay 225–350 eggs or more, and they hatch in a few days. Metamorphosis usually begins in late July to early August and lasts about 3 weeks.

SUBSPECIES There are 2 subspecies in Texas: the Eastern Cricket Frog (*Acris crepitans crepitans* Baird, 1854) and the Coastal Cricket Frog (*A. c. paludicola* Burger, Smith, and Smith, 1949).

COMMENTS AND CONSERVATION Although on endangered species lists in several states and Canada, this species is common in Texas. It is on the TPWD's Black List.

Canyon Treefrog
Hyla arenicolor, Cope, 1866

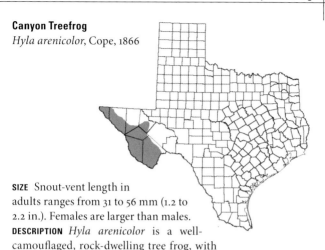

SIZE Snout-vent length in adults ranges from 31 to 56 mm (1.2 to 2.2 in.). Females are larger than males.

DESCRIPTION *Hyla arenicolor* is a well-camouflaged, rock-dwelling tree frog, with a dorsal coloration that closely matches the rocks on which it dwells. Like many tree frogs, this species can change color to more closely match the color of the background surface. Its dorsal coloration is mottled with cream, brown, greenish gray, ash gray, or brownish gray, and is often tinged with pink. The darker dorsal pattern often occurs as spots or blotches on a lighter ground, but sometimes there is little or no pattern. There is often either a dark bar or a light spot with dark edges below the eye. The underside of the frog is a cream color that grades to yellow on the hind legs, which may become orange-yellow. Toe pads are large, and the webbing on the hind feet is moderately to well developed. The skin is rather rough for a tree frog, giving the frog a distinctly toad-like appearance. The throat of males is dusky colored and may be gray, brown, or black. The vocal sac is weakly bilobed.

VOICE The call has been described as an explosive whirring that sounds like a rivet gun, all on a single pitch. The call lasts from 0.5–3 seconds. The call is similar to that of Cope's Gray Treefrog (*Hyla chrysoscelis*), but sounds as if it were coming from within a tin can.

SIMILAR SPECIES There are no other tree frogs that occur within this species range in Texas. To the inexperienced, it could be confused with a patternless Red-spotted Toad (*Anaxyrus punc-*

Canyon Treefrog, Jeff Davis County.

tatus), but toads are much wartier and have prominent parotoid glands.

DISTRIBUTION In Texas, this species occurs as disjunct populations in the Chisos, Davis, Del Norte, and Sierra Vieja Mountains. In Texas, it apparently occurs only on igneous rocks, and is absent from limestone substrates. Outside Texas, this species ranges in mountains and plateau areas from western Colorado and southern Utah southward through Arizona and New Mexico, and into Mexico as far as northern Oaxaca. There are isolated populations on the upper Canadian and Pecos Rivers in New Mexico. They occupy elevations ranging from about sea level to around 2,990 m (9,800 ft.).

NATURAL HISTORY This frog is most commonly encountered along permanent watercourses with rocky pools. It may also occur away from these streams in outcrops of large boulders or in talus slides where moisture is present year-round in the depths of rocky crevices. Although it is essentially a frog of upland wooded areas, it may also range into semiarid grasslands and cool deserts. Principally terrestrial, it often can be found huddled in shadowed nooks on the sides of boulders or stream banks, within easy jumping distance of water. They are often found stuck to the sides of rocks and roots in the permanent water areas of canyons. Although this species is primarily nocturnal, it may be active during the heat of the day in West Texas canyons. Black-necked Garter Snakes are also extremely common in these areas, and it is expected that this snake is probably one of the major predators of these tree frogs. Predators may be deterred by this frog when it releases fluid from its vent and when it releases toxic skin secretions. Prey items for Canyon Treefrogs include small scorpions, spiders and other arachnids, centipedes, annelids, and insects. Tadpoles ingest organic matter and algae that they scrape from rock and plant surfaces.

REPRODUCTION This frog breeds from March through August, usually in association with seasonal rainstorms, especially in the drier parts of its range. Reproduction occurs in pools and rocky streams, or in rain pools on the top of rock cliffs. The eggs are deposited singly, enclosed in a jelly envelope, in quiet water or in rock-lined streams and canyon bottoms. The eggs are attached to vegetation, debris, or rocks on the bottom, but they may float on the surface.

COMMENTS AND CONSERVATION The Canyon Treefrog is on the TPWD's Black List. Although this species has a restricted range in Texas, its populations seem to be stable. The chytrid fungus (*Batrachochytrium dendrobatidis*) has been reported on this frog in Arizona populations.

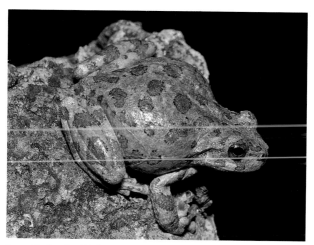

Canyon Treefrog, Presidio County.

Cope's Gray Treefrog
Hyla chrysoscelis, Cope, 1880

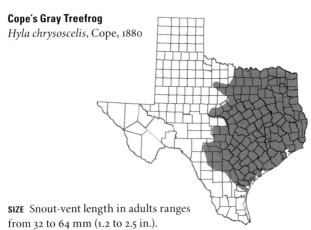

SIZE Snout-vent length in adults ranges from 32 to 64 mm (1.2 to 2.5 in.).

DESCRIPTION *Hyla chrysoscelis* is a medium-sized tree frog with skin covered in small warts. In coloration, it is typically a gray or grayish brown that is marked with large, irregular blotches of darker color. There are often dark bands on the upper surfaces of the hind limbs. Some individuals may be washed with greenish pigment at times. The coloration of an individual frog can vary dramatically, from dark gray in dormant individuals to bright green or whitish in active specimens. These color changes are the result of (at least in part) changes in the frog's activity and environment. There are 2 constant features on all gray tree frogs: a dark-edged light spot under each eye, and bold orange markings on a darker ground color on the hidden surfaces of the hind limbs. The ventral surface of this frog is white, and the toe pads are large and prominent. This species is frequently confused with *H. versicolor* because it is very similar in appearance, range, habitat, and behavior patterns (see the Similar Species section below). Much of the current literature treats one or both as if they were members of a single species, and many museums maintain only one container labeled *H. chrysoscelis/versicolor*.

VOICE The call of Cope's Gray Treefrog is a rapid, nonmusical, toad-like trill. The trill rate is 34–69 notes per second (depending upon temperature), and each individual call lasts 3–4 seconds. They may call from trees during the day, particularly in humid weather or before storms, but vigorous calling is restricted to breeding choruses.

SIMILAR SPECIES The Gray Treefrog (*Hyla versicolor*) and Cope's Gray Treefrog (*H. chrysoscelis*) are very difficult to distinguish from each other. The primary difference between these 2 closely related species is that the Cope's Gray Treefrog is a diploid animal having 2 copies of each chromosome in its cells, while the Gray Treefrog is a tetraploid animal having 4 copies of each chromosome in each cell. This difference causes the cells of the *H. versicolor* to be approximately twice as large as corresponding cells in *H. chrysoscelis*. This difference can be observed using a powerful hand lens, such as a jeweler's loupe, and is best observed by careful examination of the cells of the toe pads and vocal pouches. The larger cells of *H. versicolor* also result in this species tending to have a wartier appearance than *H. chrysoscelis*, a larger maximum size, and a slower trill rate in its calls. Although (with the proper equipment) cell size can be examined in the field, the simplest way to distinguish the 2 species is to listen to the calls of chorusing males. The call of a Cope's Gray Treefrog is a rapid nonmusical trill, and the call of the Gray Treefrog is a slower, richer, and melodic trill. The trill rate of a Cope's

Cope's Gray Treefrog, Burnet County.

Cope's Gray Treefrog, Jasper County.

Gray Treefrog is typically twice as fast as that of a Gray Tree-frog at the same temperature. The Squirrel Treefrog (*H. squirella*) may resemble this frog when it exhibits brownish coloration or when a Cope's Gray Treefrog is washed with green. But the Squirrel Treefrog is much smoother overall, lacks the dark-edged light spot beneath the eye, and lacks the mottled orange on dark ground on the hidden surfaces of the hind legs. The only other species of tree frog that may closely resemble the Cope's Gray Treefrog is the Canyon Treefrog, (*H. arenicolor*), which occurs in Texas only in the mountains of the Trans-Pecos, far to the west of the range of the Cope's Gray Treefrog.

DISTRIBUTION Because of difficulties in distinguishing *Hyla chrysoscelis* from *H. versicolor* in the field, the precise distribution of the 2 species in Texas (and throughout their ranges) has long been confused, and remains so to this day, with most field guides showing a single distribution map for both species. The composite range of these 2 species extends from southern Maine southward into northern Florida, and westward into southern Manitoba and Central Texas, with an isolated colony in New Brunswick. In Texas, these species occur throughout the eastern half of the state, ranging as far west as Uvalde County in the southern Hill Country, and Coleman and Throckmorton Counties in the Rolling Plains region. Its southernmost record is from Jim Wells County near Corpus Christi Bay. Most of the western records that have been identified to species are *H. chrysoscelis*.

NATURAL HISTORY These frogs prefer forest areas and may occur anywhere in the Texas Piney Woods, Post Oak Savannah, or Cross Timbers regions. This species penetrates drier habitats in the Rolling Plains and southern Hill Country by moving along tree-lined riparian corridors. These frogs are not often seen on the ground or at water's edge, except in breeding season, when they descend to chorus and breed. This species may be found crossing roads after heavy rainstorms in spring and summer, particularly in East Texas. This species forages high in tree canopies. At times, they can be found in rotten logs, under the bark of stumps, or in the hollows of dead trees. Cope's Gray Treefrogs are extremely well camouflaged when clinging to the bark of a rough tree trunk, and often their presence in an area is detected only when they call. Almost any small invertebrate may be taken as prey, including spiders, mites, harvestmen, and snails. Predators include various garter snakes, American Bullfrogs, and Green Frogs as well as giant water bugs. Longevity has been reported at 2–3 years, but one captive Gray Treefrog lived for more than 7 years (the species was not identified).

REPRODUCTION Breeding takes place in the spring from mid-March through June and may extend into July or even later after rainstorms. The greatest breeding activity occurs following warm rainy periods. This species prefers to breed in temporary or semipermanent pools such as ponds, flooded burrows, pits, or ditches. From 1,000 to 2,000 eggs are deposited in shallow water in small packets of 6–45 eggs, either free-floating or loosely attached to vegetation. Depending on the water temperature, *Hyla chrysoscelis* eggs hatch in 3–7 days. Metamorphosis occurs at about 6 weeks of age.

COMMENTS AND CONSERVATION This species is on the TPWD's Black List. It is one of the commonest tree frogs in Texas and so is not considered threatened or endangered. More attention is needed to distinguish the distribution of this species from that of its close relative, *Hyla versicolor*.

Green Treefrog
Hyla cinerea, (Schneider, 1799)

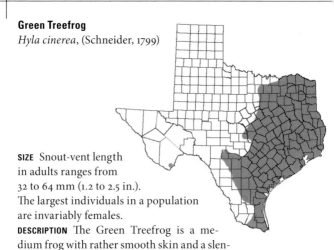

SIZE Snout-vent length in adults ranges from 32 to 64 mm (1.2 to 2.5 in.). The largest individuals in a population are invariably females.

DESCRIPTION The Green Treefrog is a medium frog with rather smooth skin and a slender, flat body. The head is rounded, with large, prominent eyes. The Green Treefrog has long slender legs with large toe pads on each foot. The vocal sac is round. The dorsal color is always some shade of green. This green color can vary from light green, when the frog is active, particularly in warm, bright conditions, to dull greenish gray, olive green, or brownish green when the frog is inactive, especially when it is hidden during a period of cool weather. Like most tree frogs, this species can change color to more closely blend in with the surface upon which they are resting. But all the reasons for their color changes are not completely understood. There typically is a white stripe present along each side of the body, and this varies in length and width from one individual to the next. In the most extensively striped individuals, the stripe may extend onto the hind legs, especially onto the shins. The lips are white, and the white color is often connected to the side stripe. Small spots of cream, yellow, or gold may be scattered randomly across the back. The irises of this species are typically golden yellow or golden brown.

VOICE The call of the Green Treefrog sounds vaguely like a honking duck. Loud and unmistakable, it is best expressed as *queenk-aueenk-queenk*, with a nasal inflection; it may be repeated as many as 75 times a minute, although the more typical rate is 30–60 notes per minute. The call has a ringing quality, and some have described it as bell-like, earning the frog such local names as "bell frog" or "cowbell frog." It typically calls from a perch 0.5–2.0 m (1.6–6.6 ft.) above permanent standing water. The calls of individual frogs can be quite loud; in fact, choruses of hundreds of males calling together have often been described as deafening. Males typically begin calling around dusk and may continue until after midnight. While choruses are mainly associated with breeding activities, they may also occur just before rain or if the weather is especially still and humid.

SIMILAR SPECIES The Squirrel Treefrog (*Hyla squirella*) matures at a much smaller size, is usually plumper, and is typically a plain green frog lacking the bold lateral stripe present in the Green Treefrog. While some Squirrel Treefrog specimens possess a light lateral stripe, the stripe's lower border is indistinct and not sharply defined, as it usually is in the Green Treefrog. The Squirrel Treefrog also has a greater capacity for color change than the Green Treefrog, with colors in a single individual ranging from pale green to a blotchy brown. Gray Treefrogs of both *H. chrysoscelis* and *H. versicolor* species may become suffused with extensive green areas, particularly on their backs, but typically possess a wartier skin than Green Treefrogs. Gray Treefrogs are thick-bodied frogs that are much stockier than Green Treefrogs. In addition, the former retain a distinctive light spot beneath each eye, typically exhibit large, blotchy areas on a dark-colored back, and lack the light stripes typical of most Green Treefrogs.

Green Treefrog, Lamar County.

Green Treefrog, Brazoria County.

DISTRIBUTION The Green Treefrog is primarily a species of the southeastern United States, ranging from the Delmarva Peninsula of Maryland, Delaware, and Virginia to the southern tip of Florida, westward through the Gulf Coastal Plain to eastern and southern Texas, and northward along the Mississippi River Valley to extreme southern Illinois, with an isolated colony in south central Missouri. In Texas, this species is found throughout the eastern third of the state, from the Oklahoma and Arkansas borders south along the Gulf of Mexico to the mouth of the Rio Grande River. There is an isolated, introduced population at Rio Grande Village in Big Bend National Park.

NATURAL HISTORY The Green Treefrog is typically found in the vicinity of permanent water. In wetter areas, it may venture afield into the surrounding woodlands, while in drier areas it is typically restricted to vegetation immediately surrounding the aquatic environment. An agile climber, it climbs nimbly through bushes, shrubs, and cattails in search of food. During the day, it is most often seen resting on a stem or leaf with its eyes closed; in such situations, its green color is highly cryptic. At dusk, this frog awakens and begins to forage in vegetation; it finds prey by sight and captures it after a short chase. In urban areas, they may be commonly seen around windows and porches, where they prey on insects that are drawn to lights. Unlike other frogs, they may be occasionally found in areas with brackish water. The Green Treefrog preys on a variety of arthropods, especially flying insect species. Prey items have been reported to include leafhoppers, grasshoppers, and spiders. While foraging, this species tends to walk rather than jump, but when threatened, it is capable of making remarkable leaps to avoid capture. Predators include snakes, other frogs, turtles, birds, mammals, fish, and invertebrates such as spiders. Because of their habit of calling in humid or rainy weather, many rural people believe these amphibians are weather prophets; however, they may call as readily before fair weather as before foul. These frogs have been reported to live up to 6 years.

REPRODUCTION Breeding calls may be heard from March to October, and the choruses may include anywhere from a few frogs to several hundred—or even several thousand—males. Breeding occurs near permanent, mostly still water, such as ponds, lakes, swamps, and other backwaters. Frogs may be seen in amplexus away from water, but actual breeding is aquatic. Eggs are deposited near the surface in small clumps of packaged jelly, often attached to floating vegetation. Clutch sizes can be quite large, to more than 2,000 eggs. Metamorphosis occurs after 55–63 days, typically from early July to October.

COMMENTS AND CONSERVATION This species is one of the most common amphibians in its range and may well have benefited from the impoundment of reservoirs throughout its range. It is on the TPWD's White List.

Squirrel Treefrog

Hyla squirella, Bosc, 1800

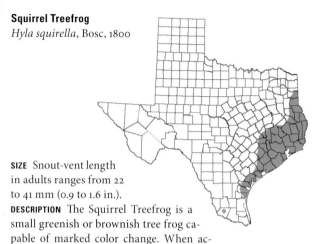

SIZE Snout-vent length in adults ranges from 22 to 41 mm (0.9 to 1.6 in.).

DESCRIPTION The Squirrel Treefrog is a small greenish or brownish tree frog capable of marked color change. When active, this species is typically bright green, often with an indistinct white stripe on the lips that extends laterally back to the forelimbs; in some specimens, the white markings may continue irregularly to the hind limbs. When resting, the same frog may be a dull brown or grayish, with scattered dark spots. The skin is noticeably smooth, the eyes are large, and the toe pads are distinct. Perhaps this frog's most distinguishing characteristic is its lack of distinguishing characteristics—until one develops a gestalt image of a Squirrel Treefrog's morphology, it is best identified by process of elimination.

VOICE The frog's breeding call, when heard from a distance, sounds like a raspy, duck-like quack. Up close, the call is a harsh trill or rasp, repeated at a rate of 1–2 calls per second, with calls repeated over a period of 10 seconds. During humid or rainy weather, outside of breeding season, this frog gives a "rain call" that closely resembles the scolding rasp of an Eastern Gray Squirrel.

SIMILAR SPECIES In Texas, this species (in its active coloration) most closely resembles a small Green Treefrog (*Hyla cinerea*). But Green Treefrogs are much larger and have proportionally longer legs and a proportionally longer body. Green Treefrogs typically have a distinct rather than an indistinct white lateral stripe. Gray Treefrogs of both species (*H. chrysoscelis* and *H. versicolor*) may be confused with a brown-phase Squirrel Treefrog, but typically have much rougher skin, a light spot below each eye, and prominent orange and brown reticulations on the hidden surfaces of the hind legs. The calls of each species readily distinguish them during the breeding season. Other similar-sized frogs include cricket frogs (*Acris*) and chorus frogs (*Pseudacris*), which either have smaller toe pads or lack toe pads, and have a more pointed snout.

DISTRIBUTION The Squirrel Treefrog is a species of the southeastern Coastal Plains, ranging from southeastern Virginia southward to the Florida Keys. Westward, it ranges throughout much of Louisiana to southeastern Texas. In Texas, it occurs along the coast to well south of Corpus Christi Bay. Except along the Texas-Louisiana border, where it ranges as far north as the Red River, this frog is not found far inland; records occur no farther inland than the vicinities of Nacogdoches and College Station. This species is occasionally transported outside its range in potted plants or on recreational vehicles. We once found a juvenile specimen in the Rio Grande Village Campground in Big Bend National Park, which had undoubtedly been inadvertently introduced in such a manner.

NATURAL HISTORY Like all tree frogs, this species is primarily nocturnal. An inhabitant of coastal marshes, it may occur in brackish water or in rainwater pools affected by salt spray. Inland, it prefers forest or woodland areas, where the shadows of trees help it retain sufficient moisture and prevent desiccation. It may be quite common in suburban and urban gardens, where it shelters in shrubs or under the eaves of houses. In dry or cold

Squirrel Treefrog, Tyler County.

Squirrel Treefrog, Washington County.

weather, this species typically becomes dormant and hides in hollow trees, under loose bark or boards in buildings, or in the branches of shrubs. During humid or rainy weather, it often gives its "rain call," and many people consider the call of this species to indicate that a storm is imminent. Prey of this species may include virtually any small arthropod, including flies, beetles, pill bugs, spiders, crickets, and ants. The frogs in turn are preyed upon by many species of snakes, birds, fish, larger frogs, turtles, small mammals, and aquatic invertebrates. Tadpole predators include dragonfly larvae, giant water bugs, and various species of predatory fish.

REPRODUCTION Breeding follows periods of heavy rain, and may occur from March to October. This species can be an explosive breeder. The males typically call from shallow standing water or the nearby shoreline. During intense rains, we have witnessed large choruses of these frogs calling directly from the asphalt of roadways. Amplexus typically occurs in water. While the eggs are laid singly, they may form clusters of up to about 1,000. These eggs typically rest on the bottoms of shallow pools. Tadpoles transform usually after 40–50 days.

COMMENTS AND CONSERVATION The Squirrel Treefrog is on the TPWD's Black List. While it is more common in states east of Texas, this frog may be locally abundant and is particularly common from the Houston area eastward to the Louisiana border.

Gray Treefrog
Hyla versicolor, LeConte, 1825

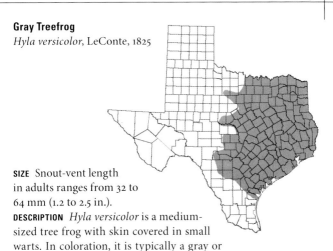

SIZE Snout-vent length in adults ranges from 32 to 64 mm (1.2 to 2.5 in.).

DESCRIPTION *Hyla versicolor* is a medium-sized tree frog with skin covered in small warts. In coloration, it is typically a gray or grayish brown that is marked with large, irregular blotches of darker color. The upper surfaces of the hind limbs often possess dark bands. Some individuals may be washed with greenish pigment (although this characteristic is more prominent in *H. chrysoscelis*). Coloration of an individual frog can vary dramatically, from dark gray in dormant individuals to whitish gray in active specimens. These color changes are the result of (at least in part) changes in the frog's activity and environment. There are 2 consistent features of all Gray Treefrogs: a dark-edged light spot under each eye, and bold orange markings on a darker ground color on the hidden surfaces of the hind limbs. The ventral surfaces of the frog are white, and the toe pads are large and prominent. This species is frequently confused with *H. chrysoscelis* because they are very similar in appearance, range, habitat, and behavior patterns (see the Similar Species section below). Much of the current literature treats both as if they were members of a single species, and many museums maintain only one container labeled *H. chrysoscelis/versicolor*.

VOICE The call of the Gray Treefrog can best be described as a musical, resonant, fluting trill that is slower than that of *H. chrysoscelis*. The trill rate is usually less than 35 pulses per minute, depending upon temperature. Calling may occur during the day in humid weather or during thunderstorms, but is most vigorous in nocturnal breeding choruses.

SIMILAR SPECIES The Gray Treefrog (*Hyla versicolor*) and Cope's Gray Treefrog (*H. chrysoscelis*) are very difficult to distinguish from each other. The primary difference between these 2 closely related species is that the Cope's Gray Treefrog is a diploid animal having 2 copies of each chromosome in its cells, while the Gray Treefrog is a tetraploid animal having 4 copies of each chromosome in each cell. This difference causes the cells of *H. versicolor* to be approximately twice as large as corresponding cells in *H. chrysoscelis*. This difference, which can be observed using a powerful hand lens such as a jeweler's loupe, is best observed by careful examination of the cells of the toe pads and vocal pouches. The larger cells of *H. versicolor* also result in this species tending to have a wartier appearance than *H. chrysoscelis*, a larger maximum size, and a slower trill rate in its calls. Although (with the proper equipment) cell size can be examined in the field, the simplest way to distinguish the 2 species is by listening to the calls of chorusing males. The call of a Cope's Gray Treefrog is a rapid nonmusical trill, and the call of the Gray Treefrog is a slower, richer, melodic trill. The trill rate of a Cope's Gray Treefrog is typically twice as fast as that of a Gray Treefrog at the same temperature. The Squirrel Treefrog (*H. squirella*) may resemble this frog when it is in its brownish coloration or when a Gray Treefrog is washed with green. But the Squirrel Treefrog is much smoother overall, lacks the

Gray Treefrog, Lamar County.

Gray Treefrog, Jasper County.

dark-edged light spot beneath the eye, and lacks the mottled orange on dark ground on the hidden surfaces of the hind legs. The only other species of tree frog that may closely resemble the Gray Treefrog is the Canyon Treefrog (*H. arenicolor*), which occurs in Texas only in the desert mountains of the Trans-Pecos, far to the west of the range of the Gray Treefrog.

DISTRIBUTION Because of the difficulties in distinguishing between *Hyla chrysoscelis* and *H. versicolor* in the field, the precise distribution of these 2 species in Texas (and throughout their range) has long been confused, and remains so to this day; most field guides simply show a single distribution map for both species. The composite range of these 2 species extends from southern Maine southward into northern Florida, and westward into southern Manitoba and Central Texas, with an isolated colony in New Brunswick. In Texas, these species occur throughout the eastern half of the state, ranging as far west as Uvalde County in the southern Hill Country, and Coleman and Throckmorton Counties in the Rolling Plains region. Its southernmost record is from Jim Wells County near Corpus Christi Bay. Most of the western records that have been identified to species are *H. chrysoscelis*.

NATURAL HISTORY These frogs prefer forest areas and may occur anywhere in the Texas Piney Woods, Post Oak Savannah, or Cross Timbers regions. This species apparently does not penetrate as far into the Rolling Plains and southern Hill Country as does the similar Cope's Gray Treefrog, which penetrates these drier habitats along tree-lined riparian corridors. These frogs are not often seen on the ground or at water's edge except in breeding season, when they descend to chorus and breed. This species may be found crossing roads after heavy rainstorms in the spring and summer, particularly in East Texas. This species forages high in tree canopies. At times, they can be found in rotten logs, under the bark of stumps, or in the hollows of dead trees. Gray Treefrogs are extremely well camouflaged when clinging to the bark of a rough tree trunk, and often their presence in an area is detected only when they call. Most any small invertebrate may be taken as prey, including spiders, mites, harvestmen, and snails. Predators include various garter snakes, American Bullfrogs, and Green Frogs as well as giant water bugs (Belostomatidae). Longevity has been reported at 2–3 years, but one captive Gray Treefrog lived for more than 7 years (the particular specimen was not identified as being either a Gray Treefrog or Cope's Gray Treefrog).

REPRODUCTION Breeding takes place in the spring, from mid-March through June, and may extend into July or even later following rainstorms. The greatest breeding activity occurs following warm rainy periods. This species prefers to breed in temporary or semipermanent pools such as ponds, flooded burrows, pits, or ditches. From 1,000 to 2,000 eggs are deposited in shallow water in small packets of 6–45 eggs, either free-floating or loosely attached to vegetation. Depending on the water temperature, *Hyla versicolor* eggs hatch in 3–7 days. Metamorphosis occurs at about 6 weeks of age.

COMMENTS AND CONSERVATION This species is on the TPWD's Black List. It is one of the most common tree frogs in Texas and so is not considered threatened or endangered.

Spotted Chorus Frog
Pseudacris clarkii, (Baird, 1854)

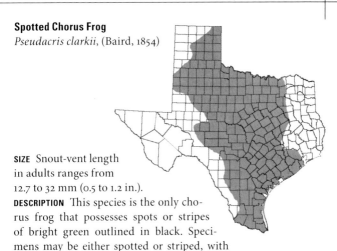

SIZE Snout-vent length in adults ranges from 12.7 to 32 mm (0.5 to 1.2 in.).

DESCRIPTION This species is the only chorus frog that possesses spots or stripes of bright green outlined in black. Specimens may be either spotted or striped, with the markings numerous or few. One consistent marking is a green triangle between the eyes, with the apex pointing backward. In spotted specimens, the spots (sometimes very numerous) are normally scattered and not arranged in rows. When stripes are present, they tend to be longitudinal, and there may be 2 or 3 stripes. A black-rimmed green stripe runs from the snout through the eye and onto the shoulder, sometimes down the side. The dorsal ground color is pale gray to gray-brown to grayish olive to ash. The belly is plain, unpatterned white. The skin is evenly textured with many small rounded warts. The toe pads are round and small.

VOICE The call is a loud rasping trill of short duration, *wrreek-wrreek-wrreek*, repeated rapidly 20 or more times. The interval between notes is about equal to the duration of the notes themselves. Individuals collect in large breeding choruses, and males call while immersed in water to the level of the throat. Often, males call from within submerged clumps of grass or other vegetation, making them difficult to observe.

SIMILAR SPECIES Strecker's Chorus Frog (*Pseudacris streckeri*) is the only other chorus frog that co-occurs with the Spotted Chorus Frog (*P. clarkii*). It is larger and more toad-like in bodily proportions, and possesses a black stripe from the snout to the shoulder, with a dark spot below the eye. Its call is described as a high-pitched, clear, bell-like *tink* rather than a rasping trill. The

Upland Chorus Frog (*P. feriarum*) exhibits brown stripes, which may be broken to some degree, but it is never highly spotted or greenish in coloration. Its call is similar, being a loud rasping trill, but has a longer duration and is not repeated rapidly.

DISTRIBUTION The Spotted Chorus Frog occurs from central Kansas to South Texas and into extreme northeastern Tamaulipas, Mexico. In Texas, it is found throughout a wide path down the middle of the state, from the central Panhandle southward to the Lower Rio Grande Valley. The Spotted Chorus Frog is relatively common in the prairie regions of Central and North Texas, occurring near sites that are seasonally flooded.

NATURAL HISTORY Very little is known about the habits of *Pseudacris clarkii*. It is primarily nocturnal and makes its home on grassland prairies and prairie islands in savannas, taking advantage of any low-lying, seasonally flooded areas for reproduction, such as semipermanent ponds, marshes, roadside ditches, grassy ponds, cattle wallows, and other temporary pools. It retreats underground during dry weather and may be found under surface rocks or other cover objects that maintain sufficient moisture. It is presumed that during the driest weather, it retreats deep underground into cracks in either the rock or soil substrate. Very little is known about the Spotted Chorus Frog's diet. Presumably, adults feed in pastures and open fields. Predators on this species are not well known, but are reported to include snakes and may include birds and other general frog predators.

Spotted Chorus Frog, Collin County.

Spotted Chorus Frog, Sterling County.

REPRODUCTION Amplexus is axillary and occurs when the frogs are submerged in temporary or semipermanent, usually fishless pools, and may last 24 hours. Breeding occurs from January through June, with the peak breeding season falling in April and May during spring rains; breeding may occur year-round in association with seasonal rainstorms in the drier, southern parts of the frog's range. Breeding may be delayed if the spring weather is dry, and may not occur at all in drought years. Up to 1,000 eggs are deposited in loose, irregular masses ranging from 3–50 eggs attached to plant stems. Eggs typically hatch in 2.5–3 days. Tadpoles metamorphose in 30–45 days.

COMMENTS AND CONSERVATION This species is on the TPWD's Black List. The Spotted Chorus Frog is subject to local declines because of habitat alteration associated with agriculture and urbanization.

Spring Peeper

Pseudacris crucifer,
(Wied-Neuwied, 1838)

SIZE These are small frogs whose snout-vent length ranges from 19 to 37 mm (0.7 to 1.5 in.). Males are usually a little larger than females.

DESCRIPTION Coloration for this species tends toward shades of yellow, pink, brown, gray, or olive. There is a dark line across the top of the head between the eyes, and a dark or indistinct *X* on the back. The legs often have dark bands. The unmarked belly is a pale cream color. The hind feet are moderately webbed, with the tips of the digits slightly expanded into pads. The males have a large vocal sac.

Spring Peeper, San Augustine County.

Spring Peeper, Jasper County.

VOICE The call is a clear high-pitched peep, repeated at intervals of 1 per second. Large choruses have been described as sounding like sleigh bells, with the sound carrying over great distances.

SIMILAR SPECIES Most other members of *Pseudacris* are distinctly striped, mottled, or spotted, with a light line along the upper lip. Gray Treefrogs (*Hyla versicolor* and *H. chrysoscelis*) have a light spot beneath the eye, are darker in color, and are much larger.

DISTRIBUTION The Spring Peeper is found from southeastern Canada southward into northern Florida and then westward into East Texas, northward through eastern Oklahoma, Kansas, northeastern Missouri, eastern Iowa, and eastern and northern Minnesota. It is associated with the deciduous and mixed forests of eastern North America.

NATURAL HISTORY The habitat of this frog is both upland and lowland woodlands within wooded marshes, ponds within brushy secondary growth or cutover woodlots, bogs and cattail wetlands, swamps, and other moist riparian undergrowth. This nocturnal frog is seldom seen except during the breeding season, but may be observed occasionally wandering about during the day when it is damp or rainy. Hibernation may occur beneath logs, bark, and fallen leaves. These frogs also have the ability to withstand subfreezing temperatures. Spring Peepers

Spring Peeper, Jasper County.

are among the earliest amphibians to emerge from hibernation. Prey items include a variety of invertebrates such as pill bugs, spiders, mites, flies, ticks, ants, beetles, springtails, and caterpillars. Tadpoles are suspension feeders, feeding on organic and inorganic material on submerged surfaces. Predators of Spring Peepers are garter snakes, giant water bugs, diving beetles, some fish, and odonate nymphs. These frogs depend on their small size, cryptic coloration, and jumping ability to avoid predation.

REPRODUCTION In Texas, they may be heard calling from November to March as long as the temperature is above freezing and it is moist. Breeding and amplexus are aquatic and can occur in any temporary to semipermanent, fishless, woodland freshwater pools. Pools such as woodland ponds, swamps, vernal pools, flooded ditches, wet meadows, cypress heads, and upland pine forests are used for breeding. The males call from thick vegetation at the water's edge. A clutch of 500–1,200 eggs are deposited singly or in clumps attached to vegetation underwater. They hatch in 3–12 days. Tadpoles will sometimes form large aggregations. Transformation occurs after 90–100 days.

COMMENTS AND CONSERVATION This species is on the TPWD's Black List. Destruction of habitat seems to be the only cause of decline in this species. It usually disappears from areas of intense human use, such as development or the clearing of forests.

Upland Chorus Frog
Pseudacris feriarum,
(Wied-Neuwied, 1838)

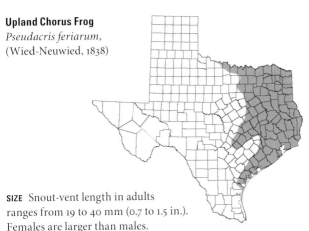

SIZE Snout-vent length in adults ranges from 19 to 40 mm (0.7 to 1.5 in.). Females are larger than males.

DESCRIPTION This is a small, slim chorus frog, with indistinct toe pads and little webbing on the feet. The dorsal coloration is highly variable, typically with ground colors ranging from gray to brown, often with a reddish or olive tint. A dark stripe extends from the snout through the eye to the groin, and contrasts with a white stripe on the upper jaw. There are usually 3 darker stripes on the back, and these are of the same

Upland Chorus Frog, amplectant pair, San Augustine County.

Upland Chorus Frog, San Augustine County.

general coloration as the ground color. In many specimens, one or more of the stripes may be broken or replaced by spots. Typically, there is a dark triangular spot present on the head. The color intensity of the stripes and the general coloration may vary with the frog's activity; for example, the same frog may appear virtually patternless when caught yet be noticeably striped a few hours later (or vice versa). The underside is whitish, yellowish, or pale olive, and is usually unmarked but may have a few dark spots on the throat and chest. The throat of the male is normally greenish yellow to dark olive, with lengthwise folds of loose skin. The vocal sac is round.

VOICE Its call is a vibrant *prreep, prreep* or *crreek, crreek* with a rising inflection, lasting about 0.5–2 seconds, with some 30–90 notes per minute. The call has sometimes been likened to the sound of running one's fingers across the fine teeth of a comb, or as a raspy rising trill. The call rises in speed and pitch as it progresses. Choruses occur night and day during the height of the breeding season, and calling typically occurs from November through May. Usually, males call from floating vegetation in open areas, but may also call from clumps of grass or beneath leaves near the edges of a pool of water.

SIMILAR SPECIES The Spring Peeper (*Pseudacris crucifer*) co-occurs with the Upland Chorus Frog, and is often found calling from the same bodies of water. It lacks the light line on the lip, is usually a bit larger and rustier in color, and usually has a dark *X* on the back. The Spotted Chorus Frog (*P. clarkii*) is similar in size, but has bright green spots or stripes.

DISTRIBUTION This species ranges from New Jersey westward to the Ohio River, then southward into extreme southeast Missouri, through Arkansas into eastern Oklahoma, and southward to Texas and the Gulf Coast.

NATURAL HISTORY A shy little frog, this is an inhabitant of grassy pools, dry grassy areas, grasslands, bogs, swamps, river bottoms, marshes, lakes, and the marshes of prairies, cultivated fields, forests, and mountains. In Texas, it is most often associated with the Piney Woods and the eastern Post Oak Savannah, being replaced by the Spotted Chorus Frog on the prairies in Central Texas. Tadpoles are herbivorous, foraging mostly on algae. Prey items of adults include small invertebrates such as flies, beetles, ants, moths, caterpillars, leafhoppers, springtails, spiders, and midges. As with other small frogs, *Pseudacris feriarum* is a preferred prey item for a large selection of predators, including snakes, birds, small mammals, other frogs, fish, cray-

Upland Chorus Frog, Houston County.

fish, turtles, and dragonfly larvae. They are reported to make an impact on insect populations where they live. These frogs may live up to 5 years in the wild.

REPRODUCTION The Upland Chorus Frog generally breeds from November to April or May, in shallow temporary pools. Occasionally, it may also use deep, more permanent water in dense woods. Although frogs may call in the day or night, most breeding occurs at night, sometimes in proximity to farms and in cities where sufficient natural habitats remain (including ephemeral breeding ponds). Eggs are laid in small loose irregular gelatinous clusters. The mass is typically 7–300 eggs (often 30–75 eggs), but a single female can contain up to 1,500 eggs. They are laid attached to vegetation and sticks in the clear, quiet water of ponds, lakes, and marshy fields. A cluster of eggs is included in an indistinct envelope. Metamorphosis generally occurs in about 2 months and depends upon water temperature.

COMMENTS AND CONSERVATION This species complex has been subdivided repeatedly by taxonomists in the last 2 decades, mostly through the use of genetic markers. This has resulted in a confusing array of species names being recognized in the scientific literature. Texas specimens have been referred to as the Western Chorus Frog (*Pseudacris triseriata*) and the Cajun Chorus Frog (*P. fouquettei*). Although heavy pesticide use may affect populations, it is still a very common frog in many areas. This species is on the TPWD's Black List.

Strecker's Chorus Frog
Pseudacris streckeri,
Wright and Wright, 1933

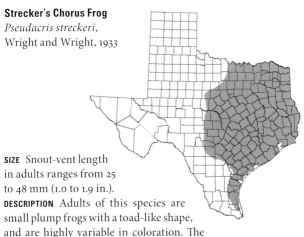

SIZE Snout-vent length in adults ranges from 25 to 48 mm (1.0 to 1.9 in.).

DESCRIPTION Adults of this species are small plump frogs with a toad-like shape, and are highly variable in coloration. The dorsal ground color ranges from pale gray to rich brown, with, in most individuals, a pattern of darker spots that may be brown, gray, or greenish, usually sharply contrasting with the ground color. Some specimens may be virtually unmarked dorsally, but these specimens typically have some spotting on the sides. Most individuals have a dark stripe that extends from the snout through the eye and onto the shoulder, and most have a dark spot below the eye. In some specimens, the dark line may extend onto the side of the body, but rarely extends to the hips. The groin is often yellowish to orange-yellow. The head is short and wide, and the forelegs are short and stout, with slightly webbed toes that have small round disks at the tips.

VOICE The call is clear and bell-like, with a single, rapidly repeated high-pitched note. In numbers, the calls merge into a single sound, much like the noise of a rapidly turning wheel or a squeaky axle. Calls are heard during and after rains from November to May, with peak activity usually in March and April.

SIMILAR SPECIES In all other chorus frogs occurring within the range of this species, there is a continuous light line along the upper lip. These other species are typically smaller and slenderer than the stocky Strecker's Chorus Frog. Toads (*Anaxyrus, Incilius, Rhinella*) and spadefoots (*Scaphiopus* and *Spea*) are larger, have rougher skin, lack toe pads, and have different coloration patterns, while similarly sized tree frogs (*Hyla*) have much longer legs and much more distinct toe pads.

DISTRIBUTION The Strecker's Chorus Frog ranges from extreme south-central Kansas southward through Texas to the Gulf of Mexico; isolated colonies have been found in South Texas, western Oklahoma, and the upper Mississippi River Valley of Arkansas, Missouri, and Illinois. In Texas, *Pseudacris streckeri* extends from the Oklahoma border southward through most of the central and eastern part of the state to the Gulf of Mexico.

NATURAL HISTORY This nocturnal and largely terrestrial frog may be found in a wide variety of habitats, but prefers well-drained, moist soils, primarily in grassy savannas and mixed woodlands, although it has also been noted to occur in rocky ravines, riparian areas of streams, lagoons, cypress swamps, marshes, and even cultivated fields. It burrows with its sturdy front feet when seeking shelter from heat or predation. While most frogs dig backward with their hind legs, *Pseudacris streckeri* is unusual in that it digs forward with its front legs and then enters the resulting hole headfirst. Strecker's Chorus Frogs prey primarily on small insects.

REPRODUCTION *Pseudacris streckeri* breeds in ephemeral ponds, including flooded fields, ditches, sloughs, and small ponds,

Strecker's Chorus Frog, Wise County.

Strecker's Chorus Frog, Lee County.

from November through midwinter into May, usually during and after rains. Peak breeding activity in Texas typically occurs in March or April. Typically, up to 700 eggs are deposited in individual egg masses of 10–100 eggs, which are attached to vegetation and twigs. They hatch within a few days. Tadpoles transform in about 60 days.

COMMENTS AND CONSERVATION This species is on the TPWD's Black List. Alarming population declines have been observed in this species. By 1976, it had disappeared completely from the Post Oak Savannah of Brazos County and from other, similar sites in eastern Texas. Because this species is terrestrial and breeds in ephemeral ponds, its froglets are more exposed to imported fire-ant predation than other species of *Pseudacris*. Because of its fossorial nature, populations of this species may be difficult to detect, especially outside the breeding season.

Mexican Treefrog
Smilisca baudinii,
(Duméril and Bibron, 1841)

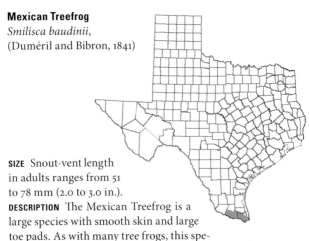

SIZE Snout-vent length
in adults ranges from 51
to 78 mm (2.0 to 3.0 in.).

DESCRIPTION The Mexican Treefrog is a
large species with smooth skin and large
toe pads. As with many tree frogs, this spe-
cies shows considerable variation, both between individuals and
within the same individual, depending upon temperature and
activity level. The most consistent markings are the light spot
beneath the eye, a dark patch or stripe running backward from
the tympanum and onto the shoulders, and a light spot at the
base of each arm. The dark patch is usually present on the same
individual throughout the extremes of its color changes. While
general coloration varies widely, this species generally ranges
from light brown or gray to grayish or brownish green to yellow or dark brown. There may also be darker blotches present on
the back and sides. Uniquely among North American tree frogs,
the inflated vocal sac of the male is paired, with each subdivi-
sion of the sac protruding downward and to the sides from beneath the throat.

VOICE The call of this species has been described as a series of
short, explosive notes sounding like *wonk-wonk-wonk* and ex-
hibiting strong carrying power. This call may sometimes be interspersed with chuckles, particularly when males are calling in
large choruses. The notes are repeated 5–12 times in 2–3 seconds,
with intervals of 2–3 seconds between calls. Males may vocalize from any body of water, but breeding choruses seem to be
most common in small temporary pools. Individuals may also
call from trees during the day in humid weather or following
rain showers. There is a distress call that may be used to attract

secondary predators, which is thought to be a defense mechanism allowing the frog to escape when its primary predator is attacked by the secondary predator.

SIMILAR SPECIES This species is most similar to the Gray Treefrogs (*Hyla chrysoscelis* and *H. versicolor*) and Canyon Treefrog (*H. arenicolor*). Neither of those species is sympatric with the Mexican Treefrog. Members of both groups are typically wartier, and the males have a single, rounded vocal pouch. Green Treefrogs (*H. cinerea*) occur sympatrically with the Mexican Treefrog, but are always bright green, always lack the dark markings on the dorsum, and also have (in males) a single, rounded vocal pouch.

DISTRIBUTION This species ranges along the Atlantic and Pacific lowlands and adjacent areas in Mexico and Central America, from extreme southern Texas (in the Lower Rio Grande Valley) on the Gulf Coast and from southern Sonora, Mexico, on the Pacific Coast southward into Costa Rica. There are historical records from Bexar and Refugio Counties, which we regard as questionable, since these records are widely separated both geographically and climatically from the remainder of the frog's recorded range.

NATURAL HISTORY This is a nocturnal frog of tropical and subtropical humid regions. It is generally considered to be a frog of the lowlands below 1,000 m (3,281 ft.), but may range into the foothills of the Sierra Madre, up to elevations of 2,000 m (6,562 ft.). They are most active after rains and may range into arid regions

Mexican Treefrog, Cameron County. Photo by Scott Wahlberg.

Mexican Treefrog, Cameron County. Photo by Scott Wahlberg.

along streams, resacas, and roadside ditches. Texas populations are thought to be small and with a fragmented distribution. During the dry season, the frog may take refuge underground, in damp tree holes, under tree bark or the outer sheaths of banana trees, or in the tops of tall palms. This species may form a cocoon in order to aestivate and avoid drought conditions. Adults are predators of invertebrates, with a preference for insects and spiders. Like other tree frogs, they are preyed upon by snakes, mammals, and birds. Aquatic insects are likely to be major predators of the tadpoles.

REPRODUCTION The Mexican Treefrog breeds with the advent of spring or summer rains, but breeding can occur year-round whenever sufficient rain falls. The timing of breeding varies geographically and corresponds closely with seasonal patterns of rainfall. Although this species may vocalize from any body of water, it seems to prefer small temporary pools. Males will call in both daytime and night, but most reproduction occurs nocturnally. Eggs are deposited in clusters and then dispersed to a surface film. Size per mass can range from 450 to 560 eggs, but females have been reported to carry more than 3,000 eggs.

COMMENTS AND CONSERVATION The Mexican Treefrog has been considered a threatened species in Texas since 1977 because of habitat fragmentation and urbanization in its limited range in the Lower Rio Grande Valley. It is not considered threatened or endangered federally. Although Texas populations are limited in distribution and abundance, this species is quite common in Mexico and Central America.

FAMILY CRAUGASTORIDAE: NORTHERN RAIN FROGS

The family Craugastoridae contains 2 genera with 115 species distributed from southern Texas to central South America. Only a single species in this family occurs as far north as Texas, reaching the Edwards Plateau and parts of the Trans-Pecos region. Eggs of members of this family are laid terrestrially or arboreally, and develop directly into miniature froglets. Members of some species in the family create a nest of foam from their skin secretions, and parental care has been observed for some species. The snout-vent length for species in this family ranges from 18 to 110 mm (0.7 to 4.3 in.). Distinguishing characteristics of the family include a first finger longer than the second, a lack of webbing between the fingers, expanded digit tips with circummarginal grooves, inner and outer metatarsal tubercles, the absence of prominent external glands on the body, the use of axillary amplexus, and distinct annuli on the tympanic membrane.

Barking Frog
Craugastor augusti,
(Duges, 1879)

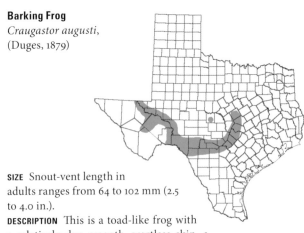

SIZE Snout-vent length in adults ranges from 64 to 102 mm (2.5 to 4.0 in.).

DESCRIPTION This is a toad-like frog with a relatively dry, smooth, wartless skin, a large head, and large front legs. There is a fold on the back of the head, dorsolateral folds on each side of the body, and a ventral disc. The color may vary from tan or brown to greenish, with dark blotches bordered by light areas on the back. Some individuals may have underlying tones of pink or reddish brown. The young have a two-toned appearance: greenish to brownish background coloration with a dark fawn to black band across

Barking Frog, adult, Real County.

Barking Frog, juvenile, Edwards County.

the middle of the back. The toes are slender and unwebbed, with small truncated toe pads and large tubercles under each joint.

VOICE The call is described as "explosive" and as resembling the bark of a dog heard at a distance, but it is more like a guttural *whurr* at close range. The single note may be repeated at regular intervals of 2–3 seconds. When females are grasped in the hand, they may make a blaring screech.

SIMILAR SPECIES The only similar species within its range is the Cliff Chirping Frog (*Syrrhophus marnockii*). The Barking Frog is much larger and has smooth skin. A juvenile Barking Frog has a white band across its otherwise dark back.

DISTRIBUTION The range extends from southeastern Arizona, southeastern New Mexico, and Central Texas (Edwards Plateau) southward into central and western Mexico to the state of Oaxaca. In Texas, it occurs in the eastern and southern edge of the Edwards Escarpment, with scattered, widely isolated localities in West Texas in Culberson, Pecos, Terrell, and Ward Counties.

NATURAL HISTORY This nocturnal, terrestrial frog is a secretive resident of limestone caves, crevices, and ledges in the Texas Hill Country, and kangaroo rat tunnels in West Texas. The frog is associated with creosote bush and mesquite trees in the western part of its range, and with maple, oak, pine, and juniper trees in the eastern part. Barking Frogs rarely venture out into the open, even during rainy periods. In the open, the frog will jump to escape, but if captured or threatened, it will inflate its body to

Barking Frog, adult, Edwards County.

several times its normal size. A Barking Frog walks with its entire body held high off the ground and is a skilled rock climber. The diet may include such invertebrates as camel crickets, grasshoppers, katydids, silverfish, and scorpions. Nothing is known about Barking Frog predators.

REPRODUCTION This frog breeds during rainy periods from late winter, usually February to May, with the peak period in March in the east and May in the west. Calling occurs from under rocks, ledges, or cracks. Eggs are deposited under rocks, in caves, in crevices, or in logs in moist areas. The tadpoles transform inside the eggs, emerging as fully developed frogs.

SUBSPECIES The subspecies found in Texas is *Craugastor augusti latrans* (Cope, 1880), the Balcones Barking Frog.

COMMENTS AND CONSERVATION This species is on the TPWD's Black List. In older literature, this frog was sometimes named *Eleutherodactylus augusti* or *Hylactophryne augusti*. We note that Barking Frogs caught near kangaroo rat mounds in West Texas and southern New Mexico secrete a noxious substance from their skin when captured, not unlike the secretions from spadefoots; these secretions can cause severe allergic reactions, such as burning or watery eyes as well as runny noses. Such secretions are absent from Barking Frogs caught along the Edwards Plateau.

FAMILY ELEUTHERODACTYLIDAE: ROBBER FROGS

The family Eleutherodactylidae contains at least 4 genera with 199 species, which occur from South Texas south into central South America, with representatives in the West Indies and southern Florida. The snout-vent length of these frogs ranges from 10.5 to 88 mm (0.4 to 3.5 in.). Small clutches of directly developing eggs—as few as a single egg in some species—are laid in terrestrial or arboreal habitats. One species within the group is ovoviparous, retaining the developing embryos within the eggs contained in the oviduct and then giving live birth to miniature froglets. Within the family, there is variation in the degree of webbing between the toes. In some species, the toes are completely unwebbed, while in others basal webbing is present, and in still others the webbing is extensive. The terminal digits are usually expanded with pads.

Rio Grande Chirping Frog
Syrrhophus cystignathoides,
(Cope, 1877)

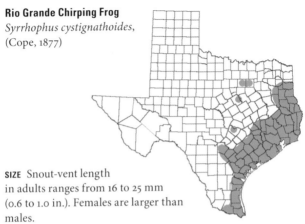

SIZE Snout-vent length
in adults ranges from 16 to 25 mm
(0.6 to 1.0 in.). Females are larger than
males.

DESCRIPTION This is a small, elongated, flattened frog with a pointed snout, well-developed forelegs, and long, slender toes with prominent tubercles and truncated toe pads. Coloration may be brown, gray, or yellowish green, with dark spots scattered across the fine granular skin, especially on the back. The legs are cross-barred, and there is a faint obscure line from the nostril through the eye. The ventral blood vessel is visible through the thin translucent skin, giving the appearance

Rio Grande Chirping Frog, Hidalgo County.

Rio Grande Chirping Frog, San Jacinto County.

of a dark line down the middle of the belly. Behavior is important in identification. This species has a habit of running instead of hopping or leaping to get away.

VOICE During rainy periods most of the year, the call is a low cricket-like chirp that is usually given erratically. When a female is present, the calls are louder and more regular.

SIMILAR SPECIES The only other species that may be similar is the Cliff Chirping Frog (*Syrrhophus marnockii*). These species may come into contact with each other on the southeastern edge of the Edwards Plateau. The Rio Grande Chirping Frog is much darker in coloration and has a less distinct dorsal pattern than the Cliff Chirping Frog.

DISTRIBUTION Native to the extreme southern tip of the Lower Rio Grande Valley in Cameron and Hidalgo Counties in Texas, it ranges south into northern Mexico. The Rio Grande Chirping Frog has been accidentally introduced into many major cities and towns from the Lower Rio Grande Valley along the coast to the Sabine River on the Louisiana border. They are also showing up inland as far north as Temple and Dallas/Fort Worth.

NATURAL HISTORY This nocturnal, terrestrial frog prefers to run rather than leap or hop, enabling it to dart under cover quickly. During the day, these frogs can be found under rocks, boards, flowerpots, debris, and other objects near palm groves, thickets, roadside ditches, resacas, creeks, and especially well-watered lawns. At night they can be seen or heard calling near their preferred habitats, especially after rain or lawn watering. This species is heard more often than it is seen. Information about prey is limited, but one individual regurgitated cockroach eggs. Predator information has not been documented.

REPRODUCTION Breeding primarily occurs in the spring, generally in April and May, but breeding calls may be heard during any of the warm months when irrigating for farming or watering of lawns occurs. Breeding is terrestrial. The females generally deposit 5–13 large eggs just under the soil surface. Direct development occurs within the egg, and tiny froglets emerge in 14–16 days.

SUBSPECIES The subspecies found in Texas is *Syrrhophus cystignathoides campi* Stejneger, 1915, the Rio Grande Chirping Frog.

COMMENTS AND CONSERVATION The Rio Grande Chirping Frog is relatively abundant in the Lower Rio Grande Valley and in its newly populated areas throughout the southern, central, and eastern parts of the state. It seems to thrive well in the midst of civilization. It is on the TPWD's Black List. They are expanding their presence in the state via the commercial distribution of plants, since it hides undetected in potting soils.

Spotted Chirping Frog
Syrrhophus guttilatus,
(Cope, 1879)

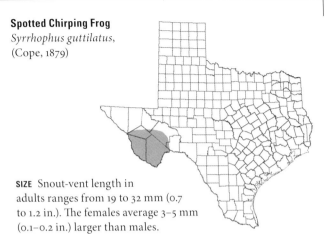

SIZE Snout-vent length in adults ranges from 19 to 32 mm (0.7 to 1.2 in.). The females average 3–5 mm (0.1–0.2 in.) larger than males.

DESCRIPTION This small frog is very similar in appearance to the Cliff Chirping Frog. This frog has a large broad flattened head and a flattened, smooth-skinned body. The toes have tubercles and truncated pads that are more prominent on the front legs than on the rear legs. Its background color ranges from yellowish to brownish, with a dark brown, irregularly spaced reticulated pattern. The hind legs have 1 or 2 dark crossbars. A dark bar is present between the eyes.

VOICE The call is similar to that of the Cliff Chirping Frog but has a more whistling quality to its sharp, short, single note. Vo-

Spotted Chirping Frog, Presidio County.

calization occurs from crevices that are 0.2–1.3 m (0.7–4.3 ft.) above the ground.

SIMILAR SPECIES The Cliff Chirping Frog (*Syrrhophus marnockii*) and Spotted Chirping Frog (*S. guttilatus*) are very difficult to differentiate. The Spotted Chirping Frog is darker, with a more pronounced reticulated pattern, a bar between the eyes, and darker hind-leg banding.

DISTRIBUTION The distribution of this frog in Texas ranges from the Big Bend region of Brewster, Presidio, and Pecos Counties southward into northern Mexico, from southeastern Coahuila to Guanajuato. These frogs occur in isolated colonies throughout their range in Texas. The occurrence of *Syrrhophus* in the Davis Mountains may represent the eastern species (*S. marnockii*) or even a new species.

NATURAL HISTORY In the mountain islands of the Chihuahuan Desert, they have been found or heard near springs, canyons, and rocky outcrops as well as along bluffs, caves, and man-made

Spotted Chirping Frog, Presidio County.

rock walls in oak-juniper woodlands. Along the Rio Grande, they have been found in mines and caves, and along road cuts and limestone bluffs. They are nocturnal and terrestrial, but can be located by day by looking under rocks, leaf litter, and debris near favorable habitat. Instead of leaping or hopping, the Spotted Chirping Frog often runs when disturbed. Stomach content analysis indicates that the frog feeds on termites, ants, beetles, and isopods. Predator information has not been documented.

REPRODUCTION Very little is known about the breeding behavior of this species. Breeding and calling are coincident with summer rains in June and July, and continue for a few weeks. Fewer than 15 eggs are laid within cracks and crevices. Larvae develop within the egg by means of direct development and hatch as fully developed froglets.

COMMENTS AND CONSERVATION The Spotted Chirping Frog is on the TPWD's Black List. They are relatively common throughout their range. Populations of this species occurring in Texas differ substantially from many populations in Mexico that are considered to be this species also, and more than one species may be recognized in the future.

Cliff Chirping Frog

Syrrhophus marnockii,
Cope, 1878

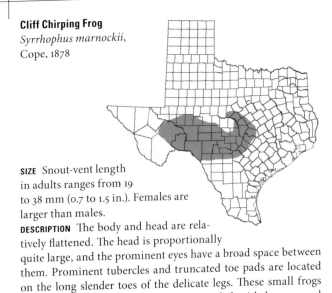

SIZE Snout-vent length
in adults ranges from 19
to 38 mm (0.7 to 1.5 in.). Females are
larger than males.

DESCRIPTION The body and head are relatively flattened. The head is proportionally
quite large, and the prominent eyes have a broad space between
them. Prominent tubercles and truncated toe pads are located
on the long slender toes of the delicate legs. These small frogs
have smooth skin that is greenish, mottled with brown, and
interspersed with light flecks. The hind legs have crossbands.

Cliff Chirping Frog, Edwards County.

These frogs are similar in appearance to Spotted Chirping Frogs (*Syrrhophus guttilatus*).

VOICE Calls can be heard throughout most of the year and are reminiscent of the chirp of a cricket, consisting of 1 or 2 brief notes, sometimes followed by a trill of 2 or 3 notes. The mating call is given only when a female is present, and is similar but clearer in tone and sharper than the advertisement call.

SIMILAR SPECIES The Green Toad (*Anaxyrus debilis*) is greener and has warty skin. The Cliff Chirping Frog (*Syrrhophus marnockii*) and Spotted Chirping Frog (*S. guttilatus*) are very difficult to differentiate. The Spotted Chirping Frog is darker, with a more pronounced reticulated pattern, a bar between the eyes, and darker hind-leg banding.

DISTRIBUTION This species ranges from south-central Texas (Edwards Plateau) westward to the Rio Grande and into the eastern Trans-Pecos region.

NATURAL HISTORY The Cliff Chirping Frog is terrestrial, nocturnal, and most active just after dusk, especially following summer and fall rain showers. Males may vocalize all night during breeding season and even during daylight hours. It leaps and hops like other frogs, but may run when seeking shelter. These

Cliff Chirping Frog, Comal County.

frogs are found in limestone cracks and crevices, in caves, under rocks, on talus slopes, in ravines, and near streams in the juniper-oak woodlands. These frogs can be abundant in urban lawns and parks that have appropriate habitat, often hiding in the crevices of rock walls. Prey items that have been reported include ants, small beetles, camel crickets, termites, and small spiders. Snakes and large wolf spiders have been known to prey on Cliff Chirping Frogs.

REPRODUCTION Reproduction is terrestrial, and breeding takes place from February to December, with the peak occurring in April or May. Breeding usually coincides with spring, summer, and fall rains. Clutches of 8–20 eggs may be laid 1–3 times per year. Tadpoles metamorphose by means of direct development and exit the eggs as fully formed froglets.

COMMENTS AND CONSERVATION The Cliff Chirping Frog is on the TPWD's Black List. They are relatively abundant in most areas, including large cities. There is much debate and confusion about the distribution and occurrence of chirping frogs in the Big Bend and Davis Mountain region of the state. Much more research on this group needs to be carried out in that part of the state.

FAMILY LEPTODACTYLIDAE: NEOTROPICAL GRASS FROGS

This Neotropical family ranges from North America southward into South America and contains more than 100 species in more than 4 genera. Only 1 species reaches as far north as South Texas. This family includes members that are fossorial, terrestrial, arboreal, or fully aquatic. Some are very small, while others can be quite large. All members of this family develop from aquatic tadpoles; some deposit their eggs in water, while others build a foam nest.

Mexican White-lipped Frog
Leptodactylus fragilis,
(Brocchi, 1877)

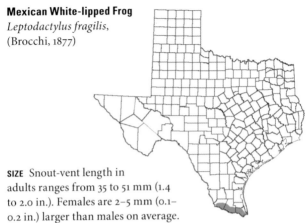

SIZE Snout-vent length in adults ranges from 35 to 51 mm (1.4 to 2.0 in.). Females are 2–5 mm (0.1–0.2 in.) larger than males on average.

DESCRIPTION This is a frog with a pointed head and prominent eyes. The Mexican White-lipped Frog has long toes, no webbing between the toes, and no toe pads. Dorsolateral folds are located low on each side of the back, and a distinct ventral disc is present on the belly. The background color can be gray, brown, or chocolate brown, with irregularly spaced, sized, and numbered dark dorsal spots. A white to cream-colored line occurs along the upper lip.

VOICE The call consists of 2 notes, with the second note rising in pitch at the end. The call is delivered in a throaty tone that sounds something like *woo-ock, woo-ock* repeated over and over.

SIMILAR SPECIES The Spotted Chorus Frog (*Pseudacris clarkii*) does not have dorsolateral folds, lacks the ventral disc, is much smaller, and has green spots.

DISTRIBUTION This is a species found as far south as Venezuela and as far north as Cameron, Hidalgo, and Starr Counties of extreme southern Texas.

NATURAL HISTORY These nocturnal frogs are found in a variety of moist habitats, such as cultivated fields, roadside ditches, irrigated fields, moist meadows, drains, potholes, oxbow lakes, resacas, low grasslands, and runoff areas. They burrow in damp soil, hiding during the day, and forage in the open at night. Males call from cavities beneath clumps of grass or dirt clods, or in small depressions as much as 8 cm (3 in.) deep.

Mexican White-lipped Frog, Tamaulipas, Mexico. Photo by Tim Burkhardt.

REPRODUCTION With the onset of heavy rains during the rainy season (late spring), breeding begins when small pools fill with water, forming the right environment for nests. Males usually call from under clumps of grass or dirt clods, or from small depressions. The small brooding depressions found under rocks, logs, or debris may be excavated by males to catch rain into which females will deposit eggs. Adults create foam nests from their body secretions by repeatedly whipping their legs on each other's bodies. The eggs are laid in the foam. The foamy nest helps preserve the eggs from desiccation until the next rain. Development occurs in the liquefied center of the foam nest until rains enable them to swim into nearby pools of water. It takes 30–35 days for development in Texas.

COMMENTS AND CONSERVATION The Mexican White-lipped Frog is listed as threatened, and is therefore protected by Texas law. Historically found in only 3 counties in South Texas, this frog may have declined through the dispersal of organophosphate chemicals in the Rio Grande Valley. Some older references use the name *Leptodactylus labialis* instead of *L. fragilis*.

FAMILY MICROHYLIDAE: NARROW-MOUTHED TOADS

This is a diverse family of frogs with about 407 species in about 68 genera found in North America, Central America, South America, Africa, Madagascar, Asia, and the Indo-Australian Archipelago. In the Americas, most members of this family are tropical. The only genera that reach the United States are *Gastrophryne* (Narrow-mouthed Toads) and *Hypopachus* (Sheep Frogs). Both are found in Texas. Members of this family in the United States are predominantly secretive, terrestrial, stout-bodied, tiny-headed frogs. A fold of skin crosses the back of the head. Legs are short, and the skin is tough but smooth. The skin probably helps protect them against the ants and termites upon which they feed.

Eastern Narrow-mouthed Toad

Gastrophryne carolinensis,
(Holbrook, 1836)

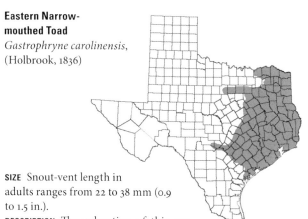

SIZE Snout-vent length in adults ranges from 22 to 38 mm (0.9 to 1.5 in.).

DESCRIPTION The coloration of this species can be gray, tan, brown, or reddish brown, with the middorsal area darker than the base color. The dark area may be flanked by broad light stripes and is frequently obscured by patches, spots, blotches, and mottling of dark or light pigment. Markings and pattern may often be completely lacking. The belly is mottled with gray, especially in the chest region. The male's throat exhibits a dark pigment. These frogs

Eastern Narrow-mouthed Toad, Harris County.

may change from one shade to another, depending on environment and activity. This toad-like frog has smooth skin and an egg-shaped body with short, stout limbs. This frog has a small head with a pointed snout and a concealed tympanum; there is a transverse fold of skin across the back of the head behind the eyes. A single enlarged tubercle, or spade, is present on the heel of each hind foot, but there is no webbing between the toes.

VOICE The call is like the bleat of a lamb, with an occasional preliminary *peep*. Some have described the 1–4 second call as having a vibrating quality, sounding something like an electric buzzer.

SIMILAR SPECIES Confusion may occur where the range of the Eastern Narrow-mouthed Toad overlaps that of the Great Plains Narrow-mouthed Toad. The Great Plains Narrow-mouthed Toad (*Gastrophryne olivacea*) has an immaculate or virtually unmarked belly, while the Eastern Narrow-mouthed Toad is strongly pigmented on the belly and chest. The Sheep Frog (*Hypopachus variolosus*) has a middorsal stripe.

DISTRIBUTION This frog ranges throughout the southeastern portion of the United States, from Maryland to the Florida Keys, and as far west as eastern Texas and Oklahoma, and north to Kentucky, southern Illinois, southern Missouri, and southeastern Nebraska.

NATURAL HISTORY This secretive, terrestrial amphibian can be found in a wide variety of habitats as long as there is shelter and moisture. They remain close to their breeding habitats of temporary pools, flooded pastures, shallow depressions, rain-filled

Eastern Narrow-mouthed Toad, Tyler County.

Eastern Narrow-mouthed Toad, Houston County.

ditches, and open grassy areas, staying sheltered under cover objects such as rocks, decaying logs, boards, mats of vegetation, animal burrows, and the bark of logs or stumps. Their habitat types include swamps, woodlands and hillsides, open woods, hardwood bottomlands, pine forests, floodplains, brackish marshes, and coastal and maritime forests. They have also been found in suburban lawns that have abundant sand and are regularly watered. Ants, termites, and small beetles are the main diet of adults. Other prey items may include snails, isopods, spiders, mites, collembolans, and small lepidopterans. Eastern Narrow-mouthed Toads can sometimes be observed at night near the opening of ant mounds, feeding on ants. Predators include water snakes, garter snakes, copperheads, cottonmouths, and egrets. The Eastern Narrow-mouthed Toad's toxic secretions can produce a burning sensation to the eyes, irritate membranes in the mouth and throat, and, therefore, may be distasteful to potential enemies. They have been reported to live more than 6 years in the wild.

REPRODUCTION The males usually call from within clumps of vegetation, chiefly in shallow water, but sometimes from deep water if it is covered by a dense floating mat of vegetation. Breeding is initiated by spring or summer rains from April through October. Females deposit a film of eggs on the surface of the water, and clutches have been reported to number from 152 to 1,600 eggs. Hatching generally occurs in 1–2 days. Metamorphosis occurs after 20–70 days.

COMMENTS AND CONSERVATION This species is on the TPWD's Black List. There have been declines in East Texas because of a loss of habitat, mainly from the construction of impoundments covering large tracts of bottomland.

Great Plains Narrow-mouthed Toad
Gastrophryne olivacea,
(Hallowell, 1857)

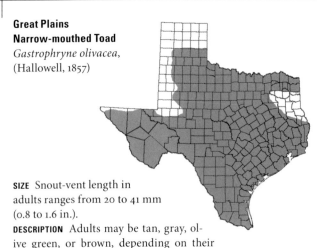

SIZE Snout-vent length in adults ranges from 20 to 41 mm (0.8 to 1.6 in.).

DESCRIPTION Adults may be tan, gray, olive green, or brown, depending on their activities and their environment. The dorsum sometimes has small scattered black spots. The ventral side is lighter than the dorsal and is unmarked, or virtually so. The young are dark brown and marked with a conspicuous dark, leaflike pattern that may occupy half the width of the back. The dark pattern disappears quickly as the frog grows. The male has a dark patch of pigment on the throat where the small vocal sac is located. The vocal sac is rounded and about the size of a pea. The body of this toad-like frog is smooth skinned, plump, and neckless; the frog has a pointed snout on its head and a broad waist. The hind legs are short and stout with fingers and toes lacking webbing. One prominent metatarsal tubercle, or spade, is located on the heel of each hind foot. A fold of skin is present across the back of the head behind the small eyes. Small tubercles are present on the lower jaw and chest. The tympanum is concealed.

VOICE The low-volume call is a short *whit* or *peep* followed by a low-pitched, nasal, insect-like buzz lasting 1–4 seconds. The pitch declines at the end of each call. At a distance, a chorus sounds like muffled, bleating sheep, but the calls are higher in pitch; nearby, it resembles the buzz of a flying bumblebee. The call may be mistaken for that of the Green Toad (*Anaxyrus debilis*).

SIMILAR SPECIES The Eastern Narrow-mouthed Toad (*Gastrophryne carolinensis*) has a strongly patterned dorsal surface, and its ventral surface is mottled with dark pigment. Sheep Frogs (*Hypopachus variolosus*) and Mexican Burrowing Toads (*Rhinophrynus dorsalis*) have a light-colored line down the center of their backs. The Sheep Frog has 2 prominent metatarsal tubercles (spades) on the heel of each hind foot.

DISTRIBUTION This small frog occurs in North America from Nebraska and central Missouri southward to the Gulf Coast, westward to southeastern Colorado, northeastern and southwestern New Mexico, and south-central Arizona, and into the lowlands of northern Mexico. This frog is found throughout Texas except for the eastern border, the western tip, and the Panhandle.

NATURAL HISTORY These nocturnal, terrestrial, and secretive frogs are a common resident of grasslands, woodlands, marshes, rocky and wooded hills and slopes, prairies, and deserts near temporary and permanent water from about sea level to around 1,250 m (4,100 ft.). During the day, these frogs seek shelter beneath rocks, boards, and other debris. They also will shelter in rodent and reptile burrows and in cracks of drying mud. These frogs are sometimes found in burrows or under rocks with tarantulas (*Aphonopelma* sp.), where they apparently live in a mutualistic relationship. The tarantula apparently protects the frog from predators, while the frog feeds on ants that may cause damage to the spider's eggs. These frogs are often heard calling

Great Plains Narrow-mouthed Toad, Fisher County.

Great Plains Narrow-mouthed Toad, Dimmit County.

immediately after spring or summer rains. In West Texas, they are most often found calling in the same bodies of water with spadefoots, Great Plains Toads, Green Toads, and Texas Toads. The adult diet consists of small invertebrates such as beetles, termites, and, especially, ants. Predators include snakes, American Bullfrogs, leopard frogs, egrets, and shrews. When escaping, they run on their short limbs or make short, rapid hops and change direction frequently. The Great Plains Narrow-mouthed Toad accumulates toxins in its body because of its diet of ants, making their skin-gland secretions distasteful to potential enemies. They have been documented to live as long as 8 years in the wild.

REPRODUCTION Breeding is usually stimulated by spring or summer rainfall occurring from March to September in temporary rain pools, cattle tanks, flooded fields, swamps, springs, and drainage or roadside ditches. Males may breed more than once a year. Researchers have reported that 500–2,000 eggs or more are deposited singly, often close together in groups forming a film on the surface of shallow water. Eggs usually hatch in 2–3 days. Tadpoles metamorphose after 28–50 days.

COMMENTS AND CONSERVATION This frog, common throughout Texas, is on the TPWD's Black List.

Sheep Frog
Hypopachus variolosus,
(Cope, 1866)

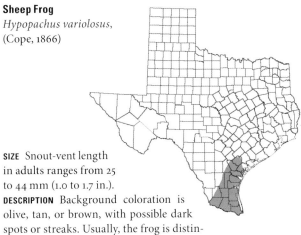

SIZE Snout-vent length in adults ranges from 25 to 44 mm (1.0 to 1.7 in.).

DESCRIPTION Background coloration is olive, tan, or brown, with possible dark spots or streaks. Usually, the frog is distinguished by a thin yellow continuous line from the snout down the length of the dorsal surface. This line is often broken into dashes or branched, and may be missing altogether. The ventral surface is gray and with some dark mottling. A thin white line running down the length of the ventral surface branches from the chest to the forelegs. Males have dark throats. Sheep Frogs

Sheep Frog, Starr County.

Sheep Frog, Starr County.

are smooth skinned and egg shaped; they have pointed snouts but no necks. A fold of skin is present across the top of the head just behind the small eyes. The toes are partially webbed, and there are 2 enlarged tubercles, or spades, on the heel of each hind leg.

VOICE The Sheep Frog's call is a clear, sheep-like, resonant bleat lasting 1.5–2.5 seconds and repeated at 10–20 second intervals.

SIMILAR SPECIES The Great Plains Narrow-mouthed Toad (*Gastrophryne olivacea*) is usually gray to light brown with no stripe down the ventral or dorsal surfaces, and it has only a single metatarsal tubercle.

DISTRIBUTION This species is found from extreme southern Texas through Mexico and Central America into Costa Rica. In South Texas, the Sheep Frog is found along the Rio Grande and north along the Gulf Coast to Aransas County and inland as far as Beeville.

NATURAL HISTORY This nocturnal, secretive frog is never found far from its breeding habitats of ponds, temporary rain pools, irrigation ditches, or roadside ditches, where it seeks protection beneath partly buried objects, such as buried tree trunks, logs, or trash-dump sites. They have also been found in mammal burrows and within the material that makes up pack rat middens. These breeding areas are found within habitats that comprise subtropical thorn scrub, savannas, open woodlands, and short-grass pasturelands. The diet of Sheep Frog adults is highly specialized for ants and termites. Predators include birds and snakes; ribbon snakes have been observed feeding on Sheep Frog tadpoles.

REPRODUCTION Breeding occurs from March to September, and is initiated by spring or summer rains or crop irrigation that creates temporary pools and fills ditches with water. Males usually call while free-floating in the water, but sometimes use their forelimbs for support on stems. Researchers have reported 600–700 eggs laid on the water surface in rafts loosely held together. They are deposited within 24 hours of heavy rainfall. Eggs hatch within 12–24 hours. Tadpoles metamorphose in about 30 days.

COMMENTS AND CONSERVATION This species is listed as threatened by the TPWD and protected from collection. Little is known regarding present-day abundances of this species in Texas.

FAMILY SCAPHIOPODIDAE: SPADEFOOTS

This is a small family of only 7 species found from southern Canada to southern Mexico. They are encountered from coast to coast in the United States. Four species are known to occur in Texas from 2 recognizable genera, *Scaphiopus* and *Spea*. A number of bony vertebral and cranial characters help unite members of this family. Other characteristics of this family may include catlike eyes, keratinized spades on hind feet, teeth on the upper jaw, rather smooth skin, indistinct or no parotoid glands on the head, pelvic amplexus, short legs, and stocky bodies. They are known as explosive breeders, appearing suddenly after heavy spring or summer rains to breed in temporary or semi-permanent pools; larval development is very rapid. Spadefoots have skin secretions that may cause strong allergic reactions, such as violent sneezing, discharge of mucus from the nose, and watering of the eyes.

Couch's Spadefoot
Scaphiopus couchii,
Baird, 1854

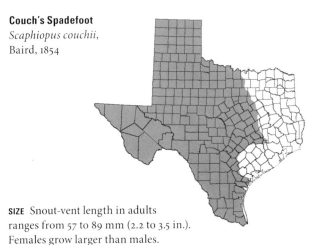

SIZE Snout-vent length in adults ranges from 57 to 89 mm (2.2 to 3.5 in.). Females grow larger than males.

DESCRIPTION This is a stout, toad-like frog with large yellow eyes, vertically elliptical pupils, and no boss between the eyes. Males tend to be greenish or yellowish in color, while in females, the yellowish ground color is overlaid with a bold black marbled network of lines. The ventral surface is whitish, and the throat is pale. The skin is predominately smooth, but close examination reveals it to be covered with many small warts, some light-

Couch's Spadefoot, male, Brewster County.

Couch's Spadefoot, female, Val Verde County.

colored and some quite dark. The eyes are widely separated, with the width of the eyelids about the same as, or less than, the distance between them. Limbs are toad-like, short, and powerful, with each hind foot possessing a dark, elongated, sickle-shaped "spade."

VOICE Best described as a cry or groan reminiscent of the bleat of a sheep, the call is a drawn-out *raa-ow* lasting 0.5–1.25 seconds and declining in pitch. A variation of the call can be described as a low-pitched raspy snore lasting approximately 1 second, similar to the beginning of the first call, but lacking the declining pitch at the end. Choruses can number several thousand calling males, and the sound that is produced from such large choruses can be heard more than 1.6 km (1.0 mi.) away.

SIMILAR SPECIES Plains Spadefoots and Mexican Spadefoots (*Spea bombifrons* and *Spea multiplicata*) have short, wedge-shaped spades and eyelids noticeably wider than the distance between them. Hurter's Spadefoots (*Scaphiopus hurterii*) and Plains Spadefoots have a boss between and just behind the eyes. Couch's Spadefoot is the only spadefoot that can be yellow.

DISTRIBUTION They range from Central Texas, southwestern Oklahoma, and central New Mexico southward, and from south central Arizona and southeastern California southward to the tip of Baja California and central Mexico. There is an isolated population in southeastern Colorado. In Texas it is found throughout the western two-thirds of the state.

NATURAL HISTORY This species of spadefoot inhabits well-drained soils of short-grass plains, vegetated deserts, creosote bush deserts, thorn shrub forests, mesquite and yucca savannas, and tropical deciduous forests in arid to semiarid regions. It is nocturnal and terrestrial and spends most of its life underground, venturing forth only after heavy spring or summer rains. The young rely on mammal burrows and cracks in the soil as retreats from the sun and heat. The adult digs its own underground burrow for shelter. They spend 8–10 months of each year underground in cocoons formed from several layers of shed skin and by a buildup of urea in the body. As long as the water potential is greater surrounding the body than within it, the spadefoot can absorb needed moisture. After heavy rainstorms, spadefoots can be found in roadside ditches, temporary pools, playas, tanks in rocky streambeds, pools in arroyos, and stock tanks. Couch's Spadefoots are reported to be found from sea level to around 1,710 m (5,600 ft.). Prey items of adults consist of invertebrates such as beetles, ants, grasshoppers, crickets, spiders, and winged and nymph termites. In the short time that they are aboveground, they have to eat enough food to provide reserves to last long periods of time while underground. Adults have been taken by Barn Owls, Swainson's Hawks, Prairie Rattlesnakes, garter snakes, and burrowing rodents.

REPRODUCTION Breeding takes place in temporary pools from April to September after heavy spring, summer, or fall rainfalls. These spadefoots have to dig to the surface, vocalize, breed, and lay eggs so that the larvae can metamorphose, all in a span of 8–15 days. Eggs are laid in cylindrical masses of more than 3,000 eggs, which are usually attached to underwater plant stems. The tadpoles transform in 8–16 days, depending on the duration of the temporary pool.

COMMENTS AND CONSERVATION This species is on the TPWD's White List. They are very common within proper habitats in the state, although few people have ever seen them. They have been negatively affected by the destruction of their historical habitats because of urban development and irrigated agriculture. But artificial stock tanks, roadside ditches, and railroad grades have offered new opportunities for colonization, and the frogs are probably more abundant now than in the past.

Hurter's Spadefoot
Scaphiopus hurterii,
Strecker, 1910

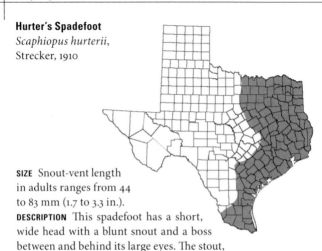

SIZE Snout-vent length
in adults ranges from 44
to 83 mm (1.7 to 3.3 in.).

DESCRIPTION This spadefoot has a short,
wide head with a blunt snout and a boss
between and behind its large eyes. The stout,
stocky body has relatively smooth skin with a scattering of small
tubercles or warts. Each hind foot has a dark, elongated, sickle-
shaped spade. The color ranges widely, from gray-green to chocolate brown, greenish brown, and almost black, with 2 irregular
light lines down the length of the back. The ventral surfaces are
white or pale gray.

VOICE An explosive bleating note lasting less than 0.5 seconds, at
intervals of about 2 seconds. These toads will call for hours, and
large choruses can be heard at great distances.

SIMILAR SPECIES Couch's Spadefoots (*Scaphiopus couchii*) and
Mexican Spadefoots (*Spea multiplicata*) have no boss between
the eyes. The Plains Spadefoot (*Spea bombifrons*) and Mexican

Hurter's Spadefoot, Hunt County.

Hurter's Spadefoot, Robertson County.

Spadefoot have a spade on each hind foot that is short, rounded, and often wedge shaped.

DISTRIBUTION They are found from central Louisiana to the Edwards Plateau of Central Texas and southward from eastern Oklahoma and western Arkansas to the Lower Rio Grande Valley.

NATURAL HISTORY Habitat for this species includes open and forested uplands, bottomlands, and cultivated farmlands that contain sandy to loamy well-draining soils. They either burrow in shallow cavities that they dig for themselves, or they may be found under logs. During breeding season, they can be encountered in a variety of temporary pools, such as temporary ponds, flooded fields, flooded roadways, roadside ditches, and borrow pits. They are found on the surface less than 29 days a year. Adults feed during their nocturnal trips, as well as at the entrance of their burrows, on a variety of arthropods. Juveniles have been collected in pine-oak forests.

REPRODUCTION Breeding can occur during any month, but usually takes place March–September after rains create temporary pools. Eggs are laid in bands of 3,000–5,500 attached to leaves, grass stems, forbs, and the stems and branches of shrubs and small trees. It takes 1–15 days for the eggs to develop and hatch. Depending on the temperature and the drying time of the pool, tadpoles transform in 14–60 days.

COMMENTS AND CONSERVATION This species is on the TPWD's Black List. Some authorities recognize the Hurter's Spadefoot (*Scaphiopus hurterii*) as a subspecies of the Eastern Spadefoot (*S. holbrookii*). Because of habitat destruction, these toads have disappeared from much of their former range.

Plains Spadefoot
Spea bombifrons, (Cope, 1863)

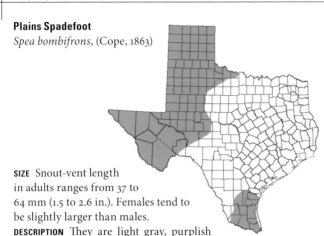

SIZE Snout-vent length
in adults ranges from 37 to
64 mm (1.5 to 2.6 in.). Females tend to
be slightly larger than males.

DESCRIPTION They are light gray, purplish
brown, or dark brown with a hint of green,
and marked with 4 light stripes, the middle pair of which are in
an hourglass pattern on the back. Dark gray or brown markings
may be found on the back, and dark pigment surrounds the tubercles. In the center of each tubercle there may also be yellow,
orange, or red specks. The ventral surfaces are white. The boss
(supported by a thickened bone) is more prominent and more

Plains Spadefoot, Fisher County.

forward than in Hurter's Spadefoot. Other characteristics include a short, stout body, smooth skin, and eyes with vertically elliptical pupils. The distance between the eyes is less than the width of the eyelids. Each hind foot has webbed toes and a single short, rounded, glossy black, wedge-shaped spade; males have keratinous pads on their thumbs.

VOICE In most populations, the frogs make a short, harsh squawk lasting 0.5–0.75 seconds and repeated over and over, but in some populations, the call is a fast trill that is shorter and more rapid than the call of the Mexican Spadefoot (*Spea multiplicata*).

SIMILAR SPECIES The Plains Spadefoot is the only Texas spadefoot that has a prominent frontal boss between the eyes. Hurter's Spadefoot (*Scaphiopus hurterii*) has a boss that is farther back between the eyes. Couch's Spadefoot (*Scaphiopus couchii*) and the Mexican Spadefoot (*Spea multiplicata*) lack a boss between the eyes. Couch's Spadefoot and Hurter's Spadefoot have an elongated spade, and the space between the eyes is about equal to the width of the eyelid.

DISTRIBUTION It ranges from southern Alberta, Saskatchewan, and southwestern Manitoba in Canada, southward into western Texas and Chihuahua in northern Mexico, and eastward following the Missouri River Valley across Missouri and Oklahoma.

Plains Spadefoot, Dawson County.

Its range skirts the eastern edge of the Rocky Mountains in the north, but in the south it is found from West Texas to eastern Arizona. In Texas, it is found in the Panhandle, the Stockton Plateau, the Trans-Pecos, and in an isolated area in the Lower Rio Grande Valley.

NATURAL HISTORY Habitats of the Plains Spadefoot consist of grasslands, sand hills, semidesert shrub, and desert shrub containing loose, well-drained soils for easy burrowing. Burrows are shallow in the summer and as deep as 4.6 m (15 ft.) in the winter, depending on moisture. In the northern and the northeastern parts of its range (in more temperate areas), it emerges with the first heavy spring rains when the temperature rises above 10°C (50°F), but in the southern and southwestern parts, it emerges during the summer rainy season. Breeding sites include cattle tanks, flooded farm fields, ditches, or playas. This spadefoot feeds nocturnally on a variety of invertebrates, including bees, wasps, ants, flies, moths, beetles, spiders, grasshoppers, termites, caterpillars, stink bugs, and other arthropods. There are both carnivorous and omnivorous tadpoles. Carnivorous tadpoles can be identified by the beaked upper jaw and the notched lower mandible. Carnivorous tadpoles feed on small invertebrates as well as other tadpoles. The omnivorous tadpoles feed on algae and detritus. Predators of spadefoot tadpoles include water beetle larvae, crayfish, and other spadefoot tadpoles. Adults may be taken by owls, hawks, snakes, and burrowing rodents.

REPRODUCTION Breeding takes place from April to August when heavy rains create temporary pools. Up to 2,000 eggs are laid in 10–250 loose cylindrical or elliptical masses attached to vegetation or other support objects in temporary or permanent pools. It takes the eggs 20 hours to hatch at 30°C (86°F).

COMMENTS AND CONSERVATION This species is on the TPWD's White List. In some parts of the Panhandle, populations have been severely reduced by agricultural practices. As with other species of spadefoots, skin secretions may cause sneezing, watering of the eyes, and mucous discharge.

Mexican Spadefoot
Spea multiplicata, (Cope, 1863)

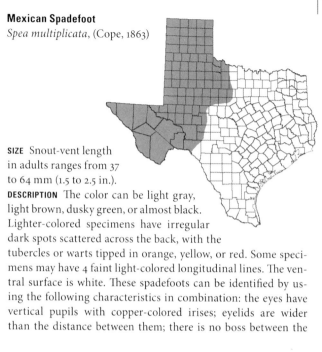

SIZE Snout-vent length in adults ranges from 37 to 64 mm (1.5 to 2.5 in.).

DESCRIPTION The color can be light gray, light brown, dusky green, or almost black. Lighter-colored specimens have irregular dark spots scattered across the back, with the tubercles or warts tipped in orange, yellow, or red. Some specimens may have 4 faint light-colored longitudinal lines. The ventral surface is white. These spadefoots can be identified by using the following characteristics in combination: the eyes have vertical pupils with copper-colored irises; eyelids are wider than the distance between them; there is no boss between the

Mexican Spadefoot, Brewster County.

Mexican Spadefoot, Crosby County.

eyes; the stout body has short legs; the smooth skin has many small flattened tubercles or warts; the hind feet have short dark wedge-shaped spades; and the skin produces an odor like fresh peanuts.

VOICE The call is a lively metallic trill lasting for about 0.75–1.5 seconds.

SIMILAR SPECIES The Plains Spadefoot (*Spea bombifrons*) and the Hurter's Spadefoot (*Scaphiopus hurterii*) have a boss between the eyes. Couch's Spadefoot (*Scaphiopus couchii*) and Hurter's Spadefoot have an elongated spade, and the space between the eyes is about equal to the width of the eyelid. In the Mexican Spadefoot there is no boss between the eyes, and the spade is short.

DISTRIBUTION Despite their standard common name, Mexican Spadefoots are found from southeastern Utah, southern Colorado, and southwestern Kansas through western Oklahoma, western Texas, New Mexico, and Arizona, and southward into central Mexico. The range in Texas includes the Panhandle, the Stockton Plateau, and the Trans-Pecos.

NATURAL HISTORY The habitat of this nocturnal spadefoot includes desert grasslands, short-grass plains, alkali flats, floodplains, creosote bush and sagebrush deserts, playas, piñon-juniper and pine-oak woodlands, and agricultural lands as long as there are temporary pools formed during heavy rains. They are absent from extreme deserts. This spadefoot is found from sea level to about 2,470 m (8,100 ft.). Spadefoots use spades on their hind feet to push dirt aside while rocking back and forth and backing into the hole. The deeper the spadefoot digs backward into the soil, the more soil falls in on top and around him. Prey items include termites, beetles, true bugs, ants, grasshoppers, crickets, spiders, and other arthropods. There are distinct carnivorous and omnivorous tadpole forms. Carnivorous tadpoles are able to transform more rapidly than omnivorous tadpoles. Larvae feed on dead organic matter, algae, crustaceans, and tadpoles. Spadefoot larvae may be preyed on by other spadefoot larvae, larval water beetles, larval Tiger Salamanders, frogs, mud turtles, grackles, and skunks.

REPRODUCTION Breeding takes place from January to August in temporary pools formed during periods of rainfall; these pools include playas, cienegas, tanks in rocky streams, temporary streams, arroyos, stock tanks, roadside and railroad ditches, and quiet streams. Typically, 300–1,000 pigmented eggs are laid in cylindrical masses in clusters of 10–42, attached to underwater vegetation or rocks. Most eggs hatch within 48 hours; however, hatching can take 0.5–6 days, depending on the temperature. Tadpoles transform in 12–44 days.

COMMENTS AND CONSERVATION This species is on the TPWD's White List. Skin secretions can cause considerable pain or discomfort if it comes into contact with the eyes, the nose, or broken skin. Habitat destruction through urbanization, water projects, and agricultural activity has negatively influenced the abundance of this spadefoot. On the other hand, the construction of stock tanks and rain-formed pools has increased their abundance in marginal habitat.

FAMILY RANIDAE: TRUE FROGS

Considered by most authorities to be "typical frogs" or "true frogs," members of the family Ranidae are normally smooth skinned and narrow waisted with long legs. The fingers are free, and the toes are joined by webs. Vocal pouches may be located at the sides of the throat or under the throat. Dorsal lateral ridges may be present or absent. These ridges are raised longitudinal folds of glandular tissue. Unlike those of other toad and frog families, some ranid females may make vocal sounds. The thumbs of breeding males are enlarged and have swollen forearms. Sexual dimorphism is evident in some species by the size of the tympanum. The family Ranidae comprises 347 species in 16 genera, which occur on all continents except Antarctica. Most lay aquatic eggs that hatch into free-swimming tadpoles. Historically, members of the genus *Lithobates* have been listed as *Rana*, and may be found under that name in older literature.

Crawfish Frog
Lithobates [Rana] areolatus,
(Baird and Girard, 1852)

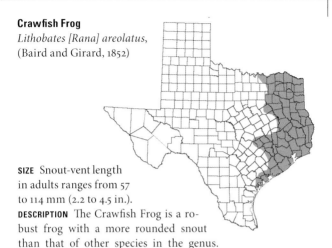

SIZE Snout-vent length in adults ranges from 57 to 114 mm (2.2 to 4.5 in.).

DESCRIPTION The Crawfish Frog is a robust frog with a more rounded snout than that of other species in the genus. Most specimens have dark dorsal spots encircled by light borders, giving the appearance of dark spots on a white to light gray background; however, this coloration is highly variable. The spots range from small to large; the largest spots are on the back, and the smaller spots are on the sides. The spots on the back sometimes are merged to form bands. The sides of the chin and throat are spotted, and the plain central throat area joins with the plain white belly. The skin on the dorsum has a rough texture. The hidden surfaces of the groin and hind legs are yellowish, and the dorsolateral folds may also have a yellowish color. This frog has short legs relative to those of other species in the family. Extensive webbing is present between the toes of its feet, with the webbing extending half the length of the longest toe.

VOICE The call is a deeply resonant, guttural trill, similar to a loud (often chuckling) deep snore. The call itself sounds something like *waaaaaaater*. Large choruses have been described as making a sound like a sty full of hogs at feeding time.

SIMILAR SPECIES The Southern Leopard Frog (*Lithobates sphenocephalus*) has a streamlined body, a very distinct dorsolateral fold, and a pointed snout. The Rio Grande Leopard Frog (*L. berlandieri*) also has a distinct dorsolateral fold and is less robust than the Crawfish Frog. Also, the coloration of both species of leopard frogs is not as reticulated as that of the Crawfish Frog. The Pickerel Frog (*L. palustris*) has regularly spaced square spots on its dorsum.

DISTRIBUTION Crawfish Frogs are found in the central region of the United States from western Indiana and southern Illinois west through central Iowa and western Missouri. They range southward into southeastern Kansas, eastern Oklahoma, eastern Texas, and then eastward to western Kentucky, central Mississippi, central Arkansas, and into northwestern Louisiana. In Texas, the Crawfish Frog is found from the central Gulf Coast eastward and northward, mainly along the Post Oak Savannah of eastern Texas, and north to the Red River. Within this range, populations are localized or disjunct in areas of suitable habitat.

NATURAL HISTORY Crawfish Frogs derived their name from their frequent use of crayfish burrows. These frogs spend the bulk of their time in crayfish burrows, mammal burrows, drainage sewers, and other types of burrows found near habitats such as wet woodlands, wooded valleys, prairies, river floodplains, pine forests, or meadows. These frogs are active on the surface only when breeding or when they are forced from their burrows during heavy rains. In Texas, the Crawfish Frog is most commonly seen in the Post Oak Savannah ecoregion; however, its secretive habits make it a difficult species to encounter. The Crawfish Frog is a generalist, eating almost anything, including insects and other frogs and reptiles, but there seems to be a preference for crayfish. Adults have been reported to take beetles, ants,

Crawfish Frog, Hunt County.

Crawfish Frog, Colorado County.

centipedes, crayfish, crickets, millipedes, and spiders. Juvenile Crawfish Frogs are likely preyed upon by animals such as water snakes and garter snakes, although no predators on adult frogs have been documented. Larvae feed mainly on phytoplankton and algae. Tadpoles are eaten by carnivorous fish and some aquatic invertebrates. The average lifespan is 5 years.

REPRODUCTION Amplexus is aquatic. Breeding occurs from February to June, but may take place year-round. Breeding in Texas is most often in February and March after heavy spring rains have filled temporary pools such as roadside ditches, pasture ponds, flooded overflows from small streams, and prairie wetlands. Egg masses are deposited in shallow water, usually near tall grass, and are sometimes attached to underwater vegetation. A clutch may contain 3,000–7,000 eggs. Hatching occurs after 7–15 days. It may take the tadpoles up to a year to metamorphose.

SUBSPECIES The subspecies found in Texas is *Lithobates areolatus areolatus* (Baird and Girard, 1852), the Southern Crawfish Frog.

COMMENTS AND CONSERVATION This species is on the TPWD's Black List. Crawfish Frogs are a rarely encountered species over most of their range and may be in decline in some areas.

Rio Grande Leopard Frog

Lithobates [Rana] berlandieri,
(Baird, 1854)

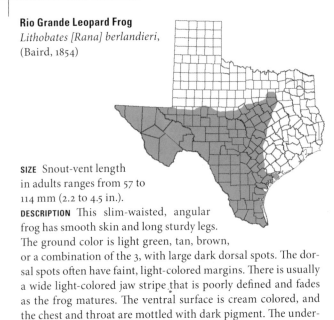

SIZE Snout-vent length
in adults ranges from 57 to
114 mm (2.2 to 4.5 in.).
DESCRIPTION This slim-waisted, angular
frog has smooth skin and long sturdy legs.
The ground color is light green, tan, brown,
or a combination of the 3, with large dark dorsal spots. The dorsal spots often have faint, light-colored margins. There is usually a wide light-colored jaw stripe that is poorly defined and fades as the frog matures. The ventral surface is cream colored, and the chest and throat are mottled with dark pigment. The underside of the hind limbs and groin area is often yellow. On the pos-

Rio Grande Leopard Frog, Brewster County.

terior surface of the thigh there are well-defined dark, contrasting reticulated areas. The base of the thumb of males is swollen and darkened. The prominent dorsolateral ridges are discontinuous and inset posteriorly. When not inflated, the male's paired vocal sacs collapse inward into a pouch forming a dark slit.

VOICE The call is a short low-pitched guttural rattle that lasts about ⅔ of a second. The call is given singly or in rapidly repeated sequences of 2–3 trills with 13 or more pulses per second.

SIMILAR SPECIES The Plains Leopard Frog (*Lithobates blairi*) has less distinct reticulations on the posterior surface of the thigh, typically with a well-defined supralabial stripe, distinct tympanum spot, and absent or faint halos surrounding the dorsal spots. The Southern Leopard Frog (*L. sphenocephalus*) is more pallid, with a distinct tympanum spot and a continuous dorsolateral ridge. The Northern Leopard Frog (*L. pipiens*) is generally much paler above and has dark spots on the posterior side of the thigh, distinct conspicuous light halos surrounding the dorsal spots, and a continuous inset dorsolateral ridge (however, *L. pipiens* has not been documented in Texas for more than 50 years, and has most likely been extirpated from the state).

DISTRIBUTION Rio Grande Leopard Frogs range from the Pecos River drainage of extreme southern New Mexico into West,

Rio Grande Leopard Frog, Edwards County.

Central, South Texas, and south along the Atlantic slope into southeastern Mexico. They can be found near sea level to around 1,520 m (5,000 ft.).

NATURAL HISTORY They are inhabitants of streams, rivers and their side pools, springs, pools along arroyos, and stock tanks within grasslands, woodlands, semiarid areas, and mountainous regions. Primarily nocturnal, this frog finds shelter from the sun and heat under rocks in thick vegetation near the water, or by hiding in mammal burrows. When approached, this nervous and excitable species usually takes to water, but may seek refuge in scrub brush or cactus or mammal burrows. Prey consists of a variety of insects and invertebrates as well as small leopard frogs. Adult frogs are preyed upon by snakes, American Bullfrogs, fish, birds, turtles, crayfish, and small mammals. Tadpoles feed on blue-green algae, green algae, inorganic particles, and diatoms. Tadpole predators include aquatic beetles, turtles, snakes, and birds. These frogs depend upon their agility and quickness to avoid predators.

REPRODUCTION Amplexus is aquatic. These frogs have been observed breeding during almost every month of the year. In the more arid parts of the range, breeding is associated with rainfall. In South Texas, they are primarily winter breeders. Clutches of 500–1,200 eggs are deposited in large masses in water and attached to submerged vegetation.

COMMENTS AND CONSERVATION This species is on the TPWD's Black List. Many of the older records refer to all leopard frogs as *Rana pipiens*, thus it can be difficult to separate these older records into actual species.

Plains Leopard Frog

Lithobates [Rana] blairi,
(Mecham, Littlejohn,
Oldham, Brown, and
Brown, 1973)

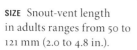

SIZE Snout-vent length
in adults ranges from 50 to
121 mm (2.0 to 4.8 in.).

DESCRIPTION This stocky pale-colored leopard frog has a beige, light brown, or dull
green dorsal surface. There are brown or olive-
green dorsal spots between and below distinct yellowish-white
dorsolateral folds. These folds are usually interrupted just past
the hip and point slightly inward. There is a white stripe running parallel to the mouth on the upper lip. The ventral surface
is white, often with fine dark stippling or mottling on the throat.
A dark spot is usually located on the pointed snout. The groin

Plains Leopard Frog, Randall County.

Plains Leopard Frog, New Mexico.

area, the lower abdomen, and the base of the thighs are often yellow. The long hind legs have long, extensively webbed toes. The roughened, deflated vocal sac of males appears as a dark lengthwise slit below the angle of the jaw. The size of the tympanum is equal to or slightly larger than the eye and is the same in both sexes.

VOICE The call is made up of 1–4 chuckle-like guttural notes, each lasting less than 1 second.

SIMILAR SPECIES The Plains Leopard Frog (*Lithobates blairi*) is stockier than the Rio Grande Leopard Frog (*L. berlandieri*) or the Southern Leopard Frog (*L. sphenocephalus*). It is also the only leopard frog in Texas with a distinct light stripe above the upper jaw.

DISTRIBUTION The Plains Leopard Frog is an inhabitant of the central and southern Great Plains. It ranges in the north from western Indiana to southeastern South Dakota and eastern Colorado, and southward from eastern New Mexico to north-central Texas. There are isolated colonies in southeastern Illinois, New Mexico, and southeastern Arizona.

NATURAL HISTORY Primary habitats of the Plains Leopard Frog include marshes, ponds, streams, ditches, river sloughs, wetlands, pools in canyons, and potholes within plains and prairies. They have also been found in oak forests, oak savannas, and oak-pine forests in the western part of their range. These frogs seek out grassy areas or other vegetative areas along the edge of their aquatic habitats. During the summer, they may be encountered some distance from water. They are primarily nocturnal, but may occasionally forage on cloudy days or during wet weather. The diet consists mostly of terrestrial organisms and may include spiders, grasshoppers, crickets, beetles, annelids, snails, and other invertebrates. Tadpoles feed on suspended aquatic matter such as algae, phytoplankton, or zooplankton. Adult predators of Plains Leopard Frogs are fish, bullfrogs, water snakes, garter snakes, raccoons, opossums, skunks, kites, and burrowing owls. To escape predators, they will leap away from, rather than toward, the water. When captured by a predator, leopard frogs will emit a loud distress call.

REPRODUCTION Reproduction and amplexus are aquatic. In the western part of their range, reproduction usually takes place after adequate summer rains. In the eastern part of their range, reproduction occurs February–October when sufficient rain falls. Typically, 200–600 (rarely up to 6,000) eggs are laid in clusters deposited in temporary or permanent pools, usually attached to plants in shallow water. Eggs hatch in 2–3 weeks. Tadpoles usually metamorphose by the middle of summer, but late-hatching tadpoles may overwinter and transform the following spring.

COMMENTS AND CONSERVATION This species is on the TPWD's Black List and is generally considered to be relatively common. In Central Texas, hybridization has occurred with the Rio Grande Leopard Frog (*Lithobates berlandieri*). Older references may use *Rana pipiens* to signify leopard frogs; however, more recent taxonomic studies have revealed that several species, including this one, were previously grouped under *R. pipiens* in Texas.

American Bullfrog
Lithobates [Rana]
catesbeianus, (Shaw, 1802)

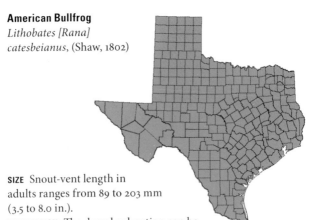

SIZE Snout-vent length in adults ranges from 89 to 203 mm (3.5 to 8.0 in.).

DESCRIPTION The dorsal coloration can be green, olive green, or brown, often grading to light green on the head. There may be a reticulated or spotted pattern of gray to brown on the dorsal background, especially in females. The upper jaw is usually light green and is unmarked. The hind legs are either banded or blotched. The ventral surface is generally white and may be mottled, often having a yellowish tinge, especially on the chin, throat, and groin. There is a fold of skin on the side of the head that extends from the eye around the tympanum to the back of the upper jaw. The tympanum of males is larger than the eye, but it is about the same size as the eye in females. Also in males, the base of the thumb is darkened and swollen. The eyes are golden in color. American Bullfrogs have a large stout broad body with large powerful hind legs; the toes are fully webbed except for the fourth digit, which extends out beyond the webbing. The skin is somewhat roughened with fine tubercles. A single internal vocal sac is present in males.

VOICE The call is a distinctively deep-pitched bellow that comes in a series of 2–4 notes; it resembles deep snores sounding like *jug-o-rum* or *br-wum*. When the frog leaps into the water to avoid capture, it will make a squawking or squeaking sound. Females may also make mating calls during breeding.

SIMILAR SPECIES The Green Frog (*Lithobates clamitans*) is similar in appearance but has dorsolateral ridges. The webbing of the toes on the hind feet of the Pig Frog (*L. grylio*) extends virtually to the tips of each toe. In American Bullfrogs, the fourth toe extends well beyond the webbing.

DISTRIBUTION Its historical range is from the Atlantic Coast westward to the Rocky Mountains and from southern Canada southward into northeastern Mexico, including all of Texas east of the Pecos River. American Bullfrogs have been extensively introduced throughout the western United States (including the Trans-Pecos region of Texas), Mexico, British Columbia, Cuba, Jamaica, Europe, and South America.

NATURAL HISTORY This is an aquatic species that is rarely seen far from permanent bodies of water. Habitats include vegetated shoals, backwaters, oxbows, farm ponds, lakes, marshes, and still water within prairies, woodlands, chaparral, forests, desert oases, and farmland. The frogs have been found from near sea level to about 2,740 m (9,000 ft.). American Bullfrogs usually hide or stay near the water's edge, resting among floating vegetation or snags. The frog's diet may include almost any organism up to the same size as itself. Smaller frogs may eat insects for the most part, while larger ones can feed on frogs, crayfish, mice, birds, and small snakes. Tadpoles feed on algae, aquatic plant material, and some invertebrates. Predators include herons, egrets, snakes, raccoons, and humans. Predation may be avoided by retreating to deeper water with a series of leaps while squawking and splashing. When seized, these frogs may make a piercing scream that can allow them to escape. American Bull-

American Bullfrog, female, Tyler County.

American Bullfrog, male, Dawson County.

frogs may be partially resistant to the venom of cottonmouths and copperheads, allowing escape.

REPRODUCTION Reproduction and amplexus is aquatic. Breeding occurs from February to August, and as late as October in the South. The most desirable males maintain their territory by puffing up their bodies, elevating themselves out of the water, and pushing, shoving, and biting competitors. Females select males with the best territories, but often swim off to another part of the pond to lay their eggs. Some females may vocalize within male choruses, thus stimulating more male-to-male competition. Clusters of eggs are laid in thin sheets, and each cluster may include 20,000–47,000 eggs. Mature females may lay as many as 3 clutches a year; eggs hatch after 3–5 days. Metamorphosis may take only a few months in the South and as long as 3 years in the North.

COMMENTS AND CONSERVATION The American Bullfrog is on the TPWD's White List. It is a common species around most permanent bodies of water, including areas where it has been introduced in the far western parts of the state. Efforts should be made to find ways to reduce or eliminate bullfrogs altogether from areas where they have been introduced, because they are voracious predators of native frog populations. In addition, the American Bullfrog is a carrier of the chytrid fungus (*Batrachochytrium dendrobatidis*).

Green Frog
*Lithobates [Rana]
clamitans*, (Latreille, 1801)

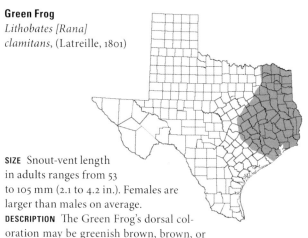

SIZE Snout-vent length
in adults ranges from 53
to 105 mm (2.1 to 4.2 in.). Females are
larger than males on average.

DESCRIPTION The Green Frog's dorsal coloration may be greenish brown, brown, or
bronze with a plain or dark-spotted pattern. There is usually
green or bronze on the sides of the head on the upper jaw. The
ventral surface is white or yellowish white with irregular dusky
lines or blotches. Young frogs are darker and more spotted above
than adults. The throat of the male is usually yellow or yellow-
orange. Prominent dorsolateral folds extend from the back of
the eye above the tympanum down the body, ending well be-

Green Frog, female, Tyler County.

Green Frog, male, Liberty County.

fore the groin. The tympanum of males is much larger than the eye, whereas in females the tympanum is the same size as the eye. The hind leg is equal to about half the body length. Toes are well webbed. The thumb base of males is darkened and swollen. When the frog calls, the paired and internal vocal sacs cause the throat and sides to expand and flatten.

VOICE The Green Frog is sometimes called the "banjo" frog because of its call. The call of this frog usually consists of a low-pitched, explosive *bung, clung,* or *c'tung* that sounds like plucking the lowest loose string of a banjo; the sound is repeated 3 or 4 times per call. When startled, the frog emits a high-pitched squawk or alert call as it leaps. When defending their territory from an intruding male, the frogs will give aggressive calls and even growl. A nonreceptive female or a male accidentally grabbed by another male may give a release call.

SIMILAR SPECIES American Bullfrogs (*Lithobates catesbeianus*) and Pig Frogs (*L. grylio*) lack dorsolateral ridges. Pickerel Frogs (*L. palustris*) and Southern Leopard Frogs (*L. sphenocephalus*) have a light line on the upper jaw, and the dorsolateral ridges extend to the groin.

DISTRIBUTION The Green Frog is a native of the eastern United States and ranges from Canada southward into north-central Florida, westward to eastern Texas, and north into the Mississippi Valley to about the mouth of the Ohio River. In the West, it has been introduced into several states and provinces, including British Columbia, Washington, Utah, and Montana.

NATURAL HISTORY Green Frogs are usually solitary, secretive, and not especially alarmed when approached. They are found in or near the permanent water of marshes, swamps, farm ponds, lakes, streams, and springs where vegetation such as cattails, rushes, sedges, and grasses are abundant. They may take shelter in logs and stumps, in crevices in limestone sinks, and in debris near water. They have also been found in brackish water. Although this species is usually crepuscular or nocturnal, it can be encountered regularly in the daytime. Green Frogs are opportunistic ambush predators. Adults feed on a variety of insects and other invertebrates as well as small vertebrates. Their diet may consist of snails, slugs, crayfish, flies, spiders, caterpillars, butterflies, moths, small snakes, and frogs. Predators have been reported on all stages of the life history of Green Frogs. Turtles have been documented eating the eggs of Green Frogs. Adults are preyed on by herons, ducks, bitterns, water snakes, garter snakes, turtles, larger frogs, and small mammals such as raccoons, otters, and minks. To escape predators, Green Frogs may emit a squawk when they jump. Their largest size is reached at 4–5 years of age, which seems to be the maximum lifespan in the wild. In captivity, Green Frogs have lived to 10 years.

REPRODUCTION Amplexus and reproduction are aquatic. The Green Frog usually breeds from March to September, with spring being the primary time. Eggs are laid in shallow water in small masses of floating clusters. Eggs are deposited near the edges of permanent quiet water and are either attached to vegetation or free-floating. Hatching occurs in 3–7 days. Metamorphosis usually takes 3 months, and the tadpoles rarely overwinter.

SUBSPECIES The subspecies found in Texas is the Bronze Frog, *Lithobates clamitans clamitans* (Latreille, 1801).

COMMENTS AND CONSERVATION This relatively common frog is on the TPWD's Black List.

Pig Frog
Lithobates [Rana] grylio,
(Stejneger, 1901)

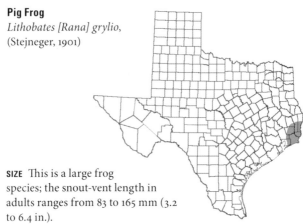

SIZE This is a large frog species; the snout-vent length in adults ranges from 83 to 165 mm (3.2 to 6.4 in.).

DESCRIPTION The Pig Frog is large and similar in appearance to the American Bullfrog, but smaller in size. It has a rather narrow, pointed head and fully webbed hind feet. The dorsal pattern and coloration can be yellow green, olive, brown, or blackish brown with scattered dark spots. The ventral surface is white or pale yellow with a netlike pattern of brown, dark gray, or black mottling on the posterior and thighs. Some specimens may have a dark pattern extending well forward on the underside of the body. The hind legs may be banded and

Pig Frog, Jefferson County.

Pig Frog, Jefferson County.

may have a light line or a row of light spots across the rear of the thighs. The tympanum is larger in males than in females. A ridge starts behind the eye and then curves above and down behind the tympanum, ending just behind the jaw. Pig Frogs have a single internal vocal pouch, but when fully extended, it has a three-part appearance.

VOICE The call, which may be heard throughout most of the year, is an abrupt low-pitched guttural grunt. Small choruses sound like a herd of foraging swine, and large groups may produce a continuous roar. When calling, males float high in shallow water.

SIMILAR SPECIES The fourth toe on the hind foot of the American Bullfrog (*Lithobates catesbeianus*) extends well beyond the webbing. The Green Frog (*L. clamitans*) has dorsolateral ridges.

DISTRIBUTION Pig Frogs are distributed along the southeastern Atlantic Coastal Plain of the United States from southern South Carolina southward into extreme southern Florida and westward to extreme southeastern Texas. It has been introduced into the Bahamas.

NATURAL HISTORY Like the American Bullfrog, the Pig Frog is primarily nocturnal and rarely active during the day except on very cloudy days. Pig Frogs are highly aquatic and may be found within emergent or floating vegetation at the edges of freshwater lakes, marshes, swamps, roadside ditches, overflowed riverbanks, and cypress ponds. During breeding season, males usually migrate to open water to call. They are wary and difficult to approach, remaining hidden in floating vegetation. Large males use calls not only during breeding season, but also to defend territory. Adults primarily prey on crawfish, but also feed on minnows, small frogs, snakes, aquatic insects, and leeches. Known predators of Pig Frogs include water snakes, cottonmouths, herons, and ibises. To defend against predators, Pig Frogs release a musty odor and a slime that is bitter to the taste.

REPRODUCTION Amplexus and reproduction are aquatic. Breeding takes place March–November, and calls can be heard throughout most of the year. Normally, 8,000–15,000 eggs per clutch are deposited as a surface film and are usually attached to vegetation. Development and hatching occurs in 2–3.5 days. They may take up to 2 years to metamorphose.

COMMENTS AND CONSERVATION The Pig Frog is on the TPWD's Black List. These frogs are relatively common and seem to be positively affected by human development. The Pig Frog is common in extreme southeastern Texas because of freshwater management practices by state and federal agencies.

Pickerel Frog
Lithobates [Rana] palustris,
(LeConte, 1825)

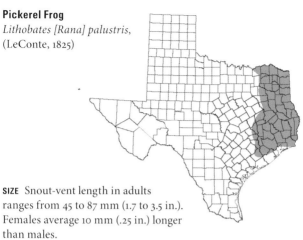

SIZE Snout-vent length in adults ranges from 45 to 87 mm (1.7 to 3.5 in.). Females average 10 mm (.25 in.) longer than males.

DESCRIPTION Pickerel Frogs are slender, smooth skinned, and relatively long legged, with distinct dorsolateral ridges. The dorsal coloration may be tan, bronze, or light brown with large, almost square, brown or reddish-brown spots arranged in 4 distinct parallel rows down the back. Usually 2 of the rows are down the middle of the back between pale dorsolateral ridges, with

Pickerel Frog, San Augustine County.

1 row to either side. The edges of the spots may be irregular, often curved, and usually outlined with white. The squares on the back may fuse to form rectangles or long longitudinal bars, giving a striped appearance. There is usually a pale line along the upper jaw and a dark streak from the nostril to the eye. The ventral surface may be white or yellow, usually with a mottling or marbling of dark lines or spots. The concealed surfaces of the hind legs are distinctly banded with bright yellow or orange. Males have stout forearms, swollen thumbs, and 2 distinct vocal sacs, one on each side below the tympanum.

VOICE One has to be fairly close to calling males to hear them. The call is a low-pitched croak or snore lasting 1–2 seconds. The call is steady, but does not carry far. Calls are often made while males are completely submerged in water, creating a gurgling-like snore.

SIMILAR SPECIES The spots on the Southern Leopard Frog (*Lithobates sphenocephalus*) are irregular and scattered, while Pickerel Frogs have spots in rows. The concealed folds of the thighs of Southern Leopard Frogs are white, while they are yellow in the Pickerel Frog.

DISTRIBUTION The Pickerel Frog ranges from southeastern Canada south to the Carolinas, and then westward through to Minnesota, south along the Mississippi River drainage, and west to eastern Oklahoma and eastern Texas. They are absent from the interior prairies and have a spotty distribution in the southern part of their U.S. range.

NATURAL HISTORY Pickerel Frogs are found in a wide variety of habitats, including streams, ponds, karst areas, wooded wet-

Pickerel Frog, Lamar County.

Pickerel Frog, Houston County.

lands, bogs, and shrubby and open meadows. It is important to note that these habitats must have relatively cool, unpolluted water with dense herbaceous vegetation. These frogs spend much of their time terrestrially traveling along stream corridors in forested areas. During breeding season, they move to aquatic habitats adjacent to their adult habitats. Very little is known about their feeding behavior. Adults may feed on insects, spiders, and other invertebrates. Tadpoles are algae and detritus feeders. Predators of tadpoles include newts, dragonfly naiads, diving beetles, and other aquatic predators. Adult predators may include bald eagles, American Bullfrogs, water snakes, herons, minks, and raccoons. Few observations have been made on antipredator mechanisms, although it is known that these defenses include toxic or distasteful skin secretions and posturing and defensive behavior. When captured, the Pickerel Frog may secrete substances that can cause irritation on any exposed skin surface and especially on the eyes when rubbed.

REPRODUCTION Reproduction and amplexus is aquatic. Pickerel Frogs breed December–May. Movements to the breeding area occur when water temperatures, air temperatures, or surface-soil temperatures reach optimum conditions. Eggs are laid most often in eutrophic zones of cool water with high levels of dissolved oxygen. Egg masses are attached to dead or living submerged vegetation near the surface in the parts of pools receiving the greatest amount of sunlight. Egg masses are spherical. Usually, eggs hatch within 10–24 days. Tadpoles metamorphose in 60–90 days after hatching.

COMMENTS AND CONSERVATION This species is on the TPWD's Black List and is uncommon over most of its range in Texas.

Southern Leopard Frog
Lithobates [Rana]
sphenocephalus,
(Cope, 1886)

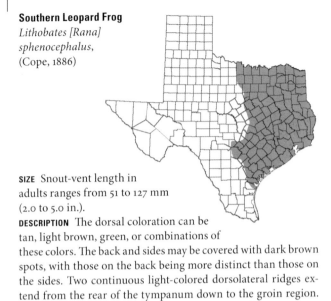

SIZE Snout-vent length in
adults ranges from 51 to 127 mm
(2.0 to 5.0 in.).

DESCRIPTION The dorsal coloration can be
tan, light brown, green, or combinations of
these colors. The back and sides may be covered with dark brown
spots, with those on the back being more distinct than those on
the sides. Two continuous light-colored dorsolateral ridges extend from the rear of the tympanum down to the groin region.
The hind legs have large brown spots that give the appearance
of bands when the frog is sitting. There usually is a light line

Southern Leopard Frog, Angelina County.

Southern Leopard Frog, Montgomery County.

along the upper jaw and a light spot on the center of the tympanum. The Southern Leopard Frog is a relatively slender leopard frog with a pointed, narrow head, long hind legs, and long toes. The 2 vocal sacs are located below the tympanum and below the jaw. When inflated, the paired vocal sacs are spherical, and form pouches when deflated.

VOICE The Southern Leopard Frog's call may include a rapid series of short deep chuckle-like croaks or a guttural trill with a rate as high as 13 pulses per second. Males call from land or while floating in shallow water.

SIMILAR SPECIES The Pickerel Frog (*Lithobates palustris*) has rectangular or square dark spots, and the concealed surfaces of the hind legs are bright yellow or orange. Crawfish Frogs (*L. areolatus*) have fat, chunky bodies with rounded, nonpointed snouts. The Rio Grande Leopard Frog (*L. berlandieri*) and the Plains Leopard Frog (*L. blairi*) have broken or discontinuous dorsolateral ridges.

DISTRIBUTION This species occupies the southeastern United States. The range extends from the Atlantic Coast of South Carolina south into Florida and westward across Kentucky to Missouri, and south into the eastern half of Oklahoma and Texas. They are found throughout the Gulf Coast.

NATURAL HISTORY In the South, the Southern Leopard Frog is active most of the year. It is found in a variety of shallow freshwater habitats, and may enter slightly brackish marshes along the coast. Its habitat may include temporary pools, cypress ponds, farm ponds, lakes, ditches, irrigation canals, stream and river edges, swamps, and marshes. These nocturnal frogs may venture away from water in summer and when the vegetation in pastures, fields, and woodlands becomes thick enough to provide shelter and shade. An adult's diet includes a variety of insects, crayfish, and other terrestrial and aquatic invertebrates. Tadpoles are mainly green algae feeders. These frogs are important food items for many predators such as herons, grackles, water snakes, racers, cottonmouths, garter snakes, small mammals, and other frogs. Large numbers have been collected for use in the bait industry and for scientific research. Many are also hunted for their hind legs. Because of this frog's alertness and quickness, it is difficult to catch either on land or in the water. Southern Leopard Frogs may escape predators by jumping into nearby water and then returning to the bank underwater at a place away from the predators' gaze.

REPRODUCTION Reproduction and amplexus are aquatic. There are 2 major breeding periods, November to March and September to October, but in the South, this frog may breed during any month of the year. The eggs are laid in shallow, still water, usually attached to submerged or emergent vegetation. Typically, 1,200–1,500 eggs or more are laid per clutch. Eggs usually hatch in 3–5 days. Tadpoles metamorphose usually in 50–75 days. Metamorphosis occurs mainly between the middle of June to the end of July.

COMMENTS AND CONSERVATION This species, which is common in Texas, is on the TPWD's Black List.

FAMILY RHINOPHRYNIDAE: BURROWING TOADS

This is a monotypic family containing a single genus with a single species, the Mexican Burrowing Toad (*Rhinophrynus dorsalis*). The family is restricted to North and Central America from extreme southern Texas southward into Honduras and Costa Rica. This is an egg-shaped, short-legged, toothless animal with a smooth, blotched skin. The body is designed for a fossorial life and the frog's termite- and ant-eating habits. The tongue is attached in the back of the mouth rather than the front, as it is in many frogs. Unlike other anurans, this frog is able to protrude its tongue like a mammal rather than flick it out in the usual toad fashion. There is a specialized hind foot for digging: the prehallux is covered with an enormous cornified "spade," and the single phalanx of the first toe is shovel-like. Although most frogs bear premaxillae and maxillae teeth, they are absent in *Rhinophrynus*, one of the few taxa in which this is the case. Although thought to be relatively common in South Texas, it is rarely seen, and then only after heavy rains.

Mexican Burrowing Toad

Rhinophrynus dorsalis,
Duméril and Bibron, 1841

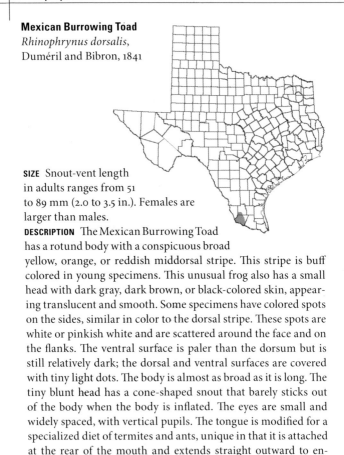

SIZE Snout-vent length
in adults ranges from 51
to 89 mm (2.0 to 3.5 in.). Females are
larger than males.

DESCRIPTION The Mexican Burrowing Toad
has a rotund body with a conspicuous broad
yellow, orange, or reddish middorsal stripe. This stripe is buff colored in young specimens. This unusual frog also has a small head with dark gray, dark brown, or black-colored skin, appearing translucent and smooth. Some specimens have colored spots on the sides, similar in color to the dorsal stripe. These spots are white or pinkish white and are scattered around the face and on the flanks. The ventral surface is paler than the dorsum but is still relatively dark; the dorsal and ventral surfaces are covered with tiny light dots. The body is almost as broad as it is long. The tiny blunt head has a cone-shaped snout that barely sticks out of the body when the body is inflated. The eyes are small and widely spaced, with vertical pupils. The tongue is modified for a specialized diet of termites and ants, unique in that it is attached at the rear of the mouth and extends straight outward to ensnare prey on its tip. The limbs are short, being better suited for burrowing than jumping. The front legs have no webbing, but the hind legs are heavily webbed to aid in swimming. Each hind foot has an elongated tubercle, or "spade," for digging. The hind legs are partially enclosed in the skin of the body. Males have a pair of internal vocal sacs. During reproduction, when they are not calling, their bodies are flaccid, and their skin seems several sizes too large for them.

SIMILAR SPECIES The general appearance is somewhat like but much larger than that of a Narrow-mouthed Toad (*Gastro-*

phryne sp.) or Sheep Frog (*Hypopachus variolosus*). The presence of the dorsal stripe and large size easily differentiates the Mexican Burrowing Toad from the Narrow-mouthed Toad. The Sheep Frog is distinguished from the Mexican Burrowing Toad by its smaller size, much lighter overall color (brownish rather than blackish), thinner yellow middorsal stripe, and the anterior attachment of its tongue.

VOICE Its call can be described as a loud, low-pitched, resonant *wh-o-o-o-oa*, lasting about 1 second and increasing in pitch toward the end. Males may call while still inside burrows, but following heavy rains, males emerge from burrows in large numbers to form massive breeding choruses that can be heard more than half a mile away.

DISTRIBUTION The Mexican Burrowing Toad is associated with lowland forests and coastal plains. Along the Gulf of Mexico, this toad is found from extreme southern Texas, south along the eastern coast of Mexico to the Yucatán Peninsula and northeastern Honduras; along the Pacific Coast, it occurs from southern Michoacan to Costa Rica.

NATURAL HISTORY These frogs are strictly fossorial and are rarely found aboveground. When active on the surface, the frogs are nocturnal, but are rarely seen other than during heavy rains, when they leave their burrows to form breeding choruses. This frog will inflate its body when approached by a would-be predator, inflating to such an extent that it almost entirely obscures its small head and limbs. The Mexican Burrowing Toad pre-

Mexican Burrowing Toad, Starr County.

Mexican Burrowing Toad, Starr County.

fers lowland areas with soil that allows for easy burrowing, using its spade-like tubercles to dig backward into the ground, hind feet first. As adults, these frogs feed primarily on termites, ants, and their associated larvae. They will also use their spade-like tubercles to dig into the mounds of their prey. Tadpoles are free swimming, and they eat phytoplankton, which is acquired through filter feeding.

REPRODUCTION Breeding is opportunistic, potentially taking place anytime following the formation of temporary pools after heavy rains. Males inflate themselves to float on the surface of temporary pools while calling in large choruses. The females emerge from their burrows to mate and lay eggs. Small clumps of thousands of eggs are laid under the water in temporary pools, where they quickly separate and float to the surface. The tadpoles tend to school, with groups numbering from 50 to several hundred individuals. These schools can contain tadpoles from different parents or earlier hatching events.

COMMENTS AND CONSERVATION The TPWD placed this species on its threatened species list in 1977. Although it is rarely seen, this species may be considered locally abundant. Choruses and specimens have been observed in recent years, especially in Starr County near the town of Rio Grande City.

APPENDIX A

Possible Additional Species for Texas

Southern Red-backed Salamander
Plethodon serratus Grobman, 1944

Mature Southern Red-backed Salamanders average 65–127 mm (2.5–5.0 in.) in total length. This is a small, rather long, slender salamander. The dorsal coloration includes striped and unstriped color morphologies that occur in the same populations. The striped phase tends to have a dark gray to dark brown ground color with a light orange to red middorsal stripe that extends from the top of the head down the length of the body to the tip of the tail. The unstriped color morph is usually the same dorsal color, and usually has small amounts of red spotting on the back, sides, and venter. This is a strictly terrestrial salamander that shows a marked preference for mesic to moderately dry hardwood forests with abundant rocks or logs and moldy leaf litter, as well as woodland debris in coniferous and hardwood forests. *Plethodon serratus* has a fragmented range in 4 disjunct regions of the southeastern United States. The closest regions to Texas are in central Louisiana and southeastern Oklahoma.

Southern Red-backed Salamander, Oklahoma.

One Southern Red-backed Salamander was collected at Fern Lake near Nacogdoches (Nacogdoches County) in 1940 and remains the only specimen encountered in Texas since its original discovery; herpetologists have searched the area over a number of years. Even though they are rare in Louisiana, a population of these salamanders exists only 137 air km (85 air miles) east of Fern Lake.

Four-toed Salamander
Hemidactylium scutatum (Temminck and Schlegel, 1838)

Adult Four-toed Salamanders measure 50–100 mm (2.0–4.0 in.) in total length. This small plethodontid adult is easily distinguished by the following characteristics: 4 toes, rather than 5, on each hind foot; a constriction around the base of the tail; and a ventral surface that is white with bold black blotches or spots. They inhabit forests that border marshes, bogs, ponds, swamps, vernal ponds, and other fish-free habitats that can serve as breeding sites. Outside the breeding season, adults in the northern part of the range are found in debris and leaf litter on the forest floor. Its general range extends from Nova Scotia southward to the Gulf of Mexico and then westward to eastern Oklahoma and Missouri, with a spotty distribution in the southern part of the range. The Four-toed Salamander is listed here as a possible resident of Texas because its range in southeastern Oklahoma falls at the Oklahoma-Texas border. This possi-

Four-toed Salamander, Tennessee.

bility is compounded because a detailed geographic analysis of this species has not been conducted.

Northern Leopard Frog
Lithobates [Rana] pipiens (Schreber, 1782)

Northern Leopard Frogs are medium-sized frogs whose snout-vent length ranges from 51 to 111 mm (2.0 to 4.4 in.). Typically, this frog is identified by a brown to green ground color with 2 or 3 rows of large round dark spots positioned between conspicuous dorsolateral ridges. The spots have light borders and may run together, and the lateral side also has rounded dark spots. The dorsolateral folds are wide and extend unbroken to the groin as a cream- or fawn-colored line. Breeding areas are associated with open-canopy forests that contain swamps, marshes, creeks, flooded ditches, small lakes, livestock tanks, or beaver ponds. In summer, they move into upland foraging areas that may include grassy areas, meadows, fields, or peat bogs. Their overwintering sites are usually close to breeding habitats. The Northern Leopard Frog can be found from southern Quebec westward to the extreme southern edge of the Northwest Territories, southward to Kentucky, westward into the Pacific Northwest, and southward into eastern Colorado and western New Mexico. The Northern Leopard Frog is listed here because, historically, it possibly occurred in Texas only in the Rio Grande Valley of the Greater El Paso area and southward to Fort Han-

Northern Leopard Frog, Nebraska.

cock. But the most recent record is more than 50 years old, and it most likely has been extirpated from the state.

Bird-voiced Treefrog
Hyla avivoca Viosca, 1928

Bird-voiced Treefrogs are medium-sized frogs with a snout-vent length ranging from 28 to 52 mm (1.1 to 2.1 in.). These gray, brown, or green tree frogs are usually found within bottomland hardwood swamps and forested floodplains in the southeastern United States, with a western population in southeastern Oklahoma, southern Arkansas, and northwestern Louisiana. It is conceivable that this species may occur in northeast Texas.

Cuban Treefrog
Osteopilus septentrionalis (Duméril and Bibron, 1841)

Cuban Treefrogs are large, with a snout-vent length of 51–152 mm (2–6 in.), and vary in color from creamy white to light brown with highlights of green, gray, beige, yellow, dark brown, or a combination of these colors. These tree frogs may reproduce in almost any watery habitat, including ponds, lakes, ditches,

Bird-voiced Treefrog, Alabama.

cisterns, and even swimming pools. They are capable of reproducing in pools of high salinity. This tree frog is native to Cuba and many other Caribbean islands. It was introduced to South Florida in the early 1920s and had become firmly established throughout peninsular Florida by the mid 1970s. Wherever this species has been introduced, it readily feeds on native frogs,

Cuban Treefrog, Florida.

which results in a population reduction of those native species. There are no documented reproducing populations in Texas, but several reputable sources have observed Cuban Treefrogs in extreme south Texas.

Greenhouse Frog
Eleutherodactylus planirostris (Cope, 1862)

Greenhouse Frogs grow to a snout-vent length of 15–30 mm (0.5–1.2 in.). These frogs have a flat elongated body with a pointed snout and reddish eyes. They lack webbing between the toes. The dorsal color of Greenhouse Frogs is brown with reddish tones, and the belly is white. There are 2 distinct color-pattern classes: striped, with longitudinal light-colored stripes, and mottled, with dark and light irregular markings. *Eleutherodactylus planirostris* is native to Cuba, but has been introduced onto many islands around the Caribbean as well as into the United States (Florida, Georgia, Mississippi, Louisiana, and Texas) and Mexico. Greenhouse Frogs are nocturnal and seek shelter under vegetation during the day. In its native habitat, the Greenhouse Frog lives in moist valleys and along small streams. The Greenhouse Frog is an invasive species in the United States. It is a recent colonizer of Galveston Island (no museum documentation). Close monitoring of adjacent coastal localities will reveal the speed with which this species colonizes new localities.

Greenhouse Frog, Florida.

APPENDIX B

Learning More about Amphibians: Resources

Texas is a great place to live if one is interested in amphibians. We suggest that readers interested in getting to know people who are also interested in amphibians contact one of the societies established just for this purpose. Some of these include the following:

Texas Herpetological Society
East Texas Herpetological Society
Austin Herpetological Society
Brazos Valley Herpetological Society
Dallas/Fort Worth Herpetological Society
South Texas Herpetology Association
West Texas Herpetological Society

International U.S.–based societies include the following:

American Society of Ichthyologists and Herpetologists (ASIH), publisher of the journal *Copeia*
Herpetologists' League (HL), publisher of the journals *Herpetologica* and *Herpetological Monographs*

Society for the Study of Amphibians and Reptiles (SSAR), publisher of the journals *Journal of Herpetology*, *Herpetological Review*, and *Catalogue of American Amphibians and Reptiles*

Texas museums include the following:

Texas Cooperative Wildlife Collection (TCWC), at Texas A&M University

Texas Natural Science Center (TNSC), at the University of Texas at Austin

Amphibian and Reptile Diversity Research Center, at the University of Texas at Arlington

Useful websites include the following:

Amphibia Web (http://amphibiaweb.org), supported by the University of California, Berkeley.

Animal Diversity Web (http://animaldiversity.org), supported by the National Science Foundation

Herps of Texas (http://www.zo.utexas.edu/research/txherps/), supported by the TNSC. The site also has a frog vocalization library.

GLOSSARY

ADPRESSED LIMBS Limbs of the animal pressed parallel to the sides of the body. The front leg is pressed backward, straight along the side, and the back leg is pressed forward, straight along the side. Adpressing works best with freshly killed or live animals, and does not work well with preserved animals.

AESTIVATE To enter a state of dormancy somewhat similar to hibernation. Aestivation usually takes place during times of heat and dryness, which corresponds often but not inevitably to the summer months.

ALLOZYME Variant form of an enzyme that is coded by different alleles at the same locus.

AMPLEXUS Sexual embrace; a process whereby a male clasps a female with his limbs before mating.

ANTERIOR Pertaining to the head or forward part of the animal.

ARBOREAL Living in trees.

AUTOTOMY Self-amputation of the tail.

AXILLA The area on the amphibian body directly under the joint where the forearm connects to the shoulder; axillary amplexus occurs when a male grasps a female from behind at this joint.

AXOLOTL Obligate neotenic salamander usually of the family Ambystomidae, but many Mole Salamander larvae in Texas are called axolotls.

BOSS Prominent ridge, protuberance, or raised knob. In toads and spadefoots, it is a protuberant part or body that forms a knob-like process on the snout or between the eyes.

CAUDAL Pertaining to the tail.

CHROMATOPHORES Pigment cells in the skin. The color of amphibians is derived from the presence of these cells. There are 3 classes of chromatophores: melanophores, containing black, brown, or red pigment; white or reflective iridophores; and xantophores, containing yellow, orange, or red pigment.

CHYTRID FUNGUS A primitive form of fungus. Most types feed on dead or rotten organic matter, but one parasitic form (*Batracochytrium dendrobatidis*, or simply Bd) can be found on the skins of amphibians, causing a disease that has produced significant population declines worldwide.

CIRRI (SINGULAR: CIRRUS) Short fleshy protuberances or nasal swellings associated with the nasolabial grooves. Cirri extend downward and transport waterborne chemicals to the nasal passage.

CITES Convention on International Trade in Endangered Species of Wild Fauna and Flora.

CLOACA (PLURAL: CLOACAE) The common channel into which the digestive, urinary, and reproductive tracts discharge their contents. The term is used for both reptiles and amphibians.

CLOACAL LIPS The smooth muscular walls of the cloaca. They control the passage of waste, eggs, and sperm caps. These lips are also referred to as the vent.

CONDYLE A rounded projection on a bone.

CONSPECIFIC Of the same species.

COSTAL FOLD A "fold" that lies between the costal grooves, actually positioned over a rib.

COSTAL GROOVES Deep vertical grooves associated with salamanders. Costal grooves are parallel on the sides and lie directly over the myosepta (connective tissue sheet) between the muscle masses. The number of costal grooves equals the number of trunk vertebrae. When counting grooves, include the groove at the rear of the front limb (in the axilla), even if it is not well developed, and both parts of a pair of grooves that may join in the groin.

CREPUSCULAR Describing animals that are primarily active during twilight, that is, at dawn and at dusk.

CREST An elevated section on the head.

CUSP A projection arising from the surface of the crown of a tooth.

DESICCATION The process of becoming extremely dry.

DERIVED SPECIES Species that has changed from an ancestral form.

DERMIS Inner layer of skin.

DETRITUS Leaf litter and other dead organic matter intermixed with soil.

DIGIT A toe or finger.

DIPLOID In vertebrates, having the usual 2 sets of chromosomes.

DIURNAL Active during daylight hours.

DORSAL (OR DORSUM) The upper surface of an organism.

DORSOLATERAL FOLDS Folds lengthwise down the sides of the body.

ECTOTHERM An organism that is dependent on an external heat source to warm its body.

EFT STAGE The terrestrial juvenile life stage of a newt.

EGG An animal reproductive body consisting of an ovum together with its nutritive and protective envelopes and having the capacity to develop into a new individual capable of independent existence.

EPHEMERAL POOL A body of water that is a wetland, spring, stream, river, pond, or lake that exists for only a short period following precipitation or snowmelt.

EPIDERMAL GLANDS Glands that are located in the epidermis (skin).

EPIDERMIS Outer skin.

EUTROPHIC ZONE Area in which the concentration of chemical nutrients (pollutants) in an ecosystem has increased to the extent that the primary productivity of the ecosystem is increased, but the oxygen content available to other organisms is decreased.

EXTIRPATION The extinction of an animal from a specific portion of its range.

FEMORAL WARTS Thigh warts.

FIMBRIAE *See* gill fimbriae.

FORELIMB Front of the furthest tip of the toes (digits) to the elbow.

FOSSORIAL Adapted for digging and living underground.

GESTALT Visual or mental representation of figures as whole forms rather than collections of simple lines and curves.

GILL CLEFT Gill slit.

GILL FIMBRIAE Hairlike extensions on the surface of the gills that facilitate gas exchange.

GILL RAKERS Cartilaginous or bony extensions near the base of the gill arch, or rachis. These processes prevent solid objects from passing through the gill clefts, thereby acting like "gill colanders."

GILL SLIT The space separating 2 adjoining gill arches. It allows water to pass from the throat to the outside of the organism.

GRANULAR GLAND A gland that produces and secretes granular material such as hormones.

GRAVID Describing the condition of reptiles and amphibians when carrying eggs internally.

GROUND COLOR The main background color of an organism.

GULAR FOLD An external fold or flap of tissue that extends across the lower throat region of salamanders.

HOLOTYPE A specimen on which the description of a new species is based.

INTEGUMENT Skin.

INTERORBITAL DISTANCE Distance between the eyes.

IRIDOPHORES White or reflective chromatophores. These cells produce brassy, silvery, golden, or shiny colors.

JUVENILE A sexually immature individual.

KARST Landscape shaped by the dissolution of a layer or layers of soluble bedrock, usually carbonate rock such as limestone or dolomite. Caves and sinkholes are a result.

KEEL A thin raised edge that runs along the dorsal surface of the tail.

LABIAL Pertaining to the lip.

LARVA (PLURAL: LARVAE) In amphibians, the gilled immature form that hatches from an egg and is fundamentally unlike the parent. It must metamorphose before assuming the morphology of the sexually mature parent. Neotenic forms never completely metamorphose into another form.

LATERAL Toward the sides of the body.

MEDIAN Midline of the body.

MELANISTIC Referring to an exceptionally dark individual or structure whose coloration is owed to the presence of large quantities of the pigment melanin.

MELANOPHORE Black, brown, grayish, or red chromatophores.

MENTAL GLANDS Secretory glands present on the chins of some male salamanders; they are used in courtship.

METAMORPH In amphibians, an individual recently transformed from a gilled larval stage to an ungilled juvenile stage.

METAPODIALS Bones between the wrist and fingers on a forelimb, or between the ankle and toes on a hind limb.

METATARSAL TUBERCLE The area of the hind limb occupied by any one of the metapodials in the hind foot. Found from the heel to the base of the recognized digits.

METAMORPHOSIS Process of transformation from a gilled larval stage to an ungilled juvenile stage.

MICROHABITAT Within an organism's overall habitat, a small area with unique characteristics.

MORPH One of a small number of variants in color or morphology that occur in a group.

MYOSEPTA Connective tissue sheet that is found between muscle masses.

NARIS (PLURAL: NARES) A nostril or external opening to the nasal passages.

NASOLABIAL GROOVE Slit-like channel that extends from the margin of the upper lip to the lateral corner of each nostril and sometimes out onto a lobe (or palp). A characteristic of the family Plethodontidae, these channels transport waterborne chemicals from the substrate to the vomeronasal organ and are important in facilitating chemically mediated behaviors. In this salamander family, the groove is present if external gills are absent, and vice versa.

NOCTURNAL Active at night.

NEOTENY Attainment of sexual maturity during the larval stage, or retention of some larval or immature characteristics in the adult phase.

NOMENCLATURE The system of scientific names applied to taxa.

OPISTHOCOELOUS Described as concave behind. Refers especially to the vertebrae in which the anterior end of the centrum is convex and the posterior is concave.

OVIDUCT A tube that transports eggs from the ovaries toward the cloaca.

OVIPOSIT To lay eggs or egg masses.

PAEDOMORPHIC Describing the phenotypic or genotypic change in which the adults of a species retain traits previously seen only in juveniles.

PAROTOID GLAND Toxic granular gland behind the eye and to the side of the head in some amphibians; in some individuals, the gland extends onto the back of the neck and body.

PERENNIBRANCH (OR PERENNIBRANCHIATE) Nontransforming, permanently gilled salamander.

PHALANX (PLURAL: PHALANGES) Finger or toe bone.

PHARYNX Throat.

PHOTOTAXIS Movement in response to stimulation by light. Can be positive (toward) or negative (away from) the light stimulus.

PLAYA Basins found in deserts and grasslands. They lack outlets and fill with water after rains to form temporary lakes.

POND-TYPE LARVA Larva having bushy gills and a prominent fin that extends forward to near the shoulder region.

POSTERIOR Pertaining to the back end or rear.

PREHALLUX An extra first toe or a rudiment of a toe.

PREMAXILLARY TEETH Teeth at the very anterior of the mouth.

PREVOMERINE TEETH Teeth that occur in a transverse row that crosses the palate near the posterior margin of the internal nares.

RACHIS (PLURAL: RACHISES) Central section of a gill that supports the finely divided gill fimbriae.

RAKERS *See* gill rakers.

RESACA A former channel of the Rio Grande; found in southern Texas and northern Mexico.

SNOUT-VENT LENGTH Distance from the tip of the snout to the posterior margin of the cloaca.

SPERM CAP Mass of seminal fluid resting upon the top of the jelly base of a spermatophore.

SPERMATHECA (PLURAL: SPERMATHECAE) Sperm-storage chamber leading off the internal lining of the female cloaca.

SPERMATOPHORE A structure deposited on substrates by courting male salamanders, typically consisting of a mass of seminal fluid and the sperm cap resting on a gelatinous base.

STREAM-TYPE LARVA Larva having reduced gills and a fin that extends forward only to the insertion of the hind limbs.

SUPRAORBITAL CREST Crest behind and above the eye.

SYMPATRIC Describing 2 or more species or groups with overlapping ranges and living in the same geographic region.

TALUS SLIDES A sloping mass of fragmented rock lying at the foot of a steep incline or mountain.

TARSUS Flat of the foot; the part of the foot of a vertebrate between the metatarsus and the leg; also the small bones that support this part of the limb.

TAXON (PLURAL: TAXA) Taxonomic name applied to a group of organisms.

TETRAPLOID In vertebrates, having 4 sets of chromosomes instead of the usual 2.

TIBIA WARTS Shin warts.

TOTAL LENGTH The distance from the tip of the snout to the tip of the tail.

TRANSFORMATION Metamorphosis; the transition from a gilled larval form to an ungilled juvenile form.

TRIPLOID In vertebrates, having 3 sets of chromosomes instead of the usual 2.

TROGLODYTE Cave-dwelling organism.

TRUNK VERTEBRAE Body vertebrae.

TWILIGHT ZONE Zone at a cave mouth where light is visible.

TYMPANUM Eardrum.

UNKEN REFLEX Passive defensive posture in which amphibians display their brightly colored inter-surfaces and/or undersurfaces as a warning to predators.

UREA Chemical compound found in urine.

UROSTYLE A bone composed of fused vertebrae in some fishes and tailless amphibians; found at the posterior end of the spinal column.

VENT In reptiles and amphibians, the external opening of the cloaca.

VENTER OR VENTRAL Belly; the lower surface of the organism.

VERNAL POND A seasonally ephemeral depression that typically fills with water during winter or early spring and dries before summer or early fall.

VESTIGIAL OVIDUCT Formerly called the Mullerian duct; found along each kidney and associated with male frogs.

VITELLINE MEMBRANE Innermost protective membrane surrounding a developing embryo.

VOMERINE TEETH Teeth positioned on the paired bones on the roof of the mouth.

VOUCHER SPECIMEN A specimen collected and preserved in a museum to document its occurrence in a state, county, or region.

WART Typically, a collection of granular cells, often toxic.

XERIC Describing a condition of little moisture.

YPSILOID CARTILAGE A reduced pectoral, or chest, girdle that supports the forelimbs and remains in a cartilaginous condition. It is used for exhalation in some salamanders.

BIBLIOGRAPHY

Bartlett, R. D., and P. P. Bartlett. 1999. *A field guide to Texas reptiles and amphibians*. Houston: Gulf.

Bartlett, R. D., and P. P. Bartlett. 2006. *Guide and reference to the amphibians of eastern and central North America (north of Mexico).* Gainesville: Univ. of Florida Press.

Behler, J. L. 1979. *The Audubon Society field guide to North American reptiles and amphibians*. New York: Knopf.

Burger, W. L., P. W. Smith, and F. E. Potter, Jr. 1950. Another neotenic *Eurycea* from the Edward's Plateau. *Proceedings of the Biological Society of Washington* 63:51–58.

Chippindale, P. T. 2000. Species boundaries and species diversity in the Central Texas hemidactyline plethodontid salamanders, genus *Eurycea*. In *The biology of plethodontid salamanders*, ed. R. Bruce, L. Houck, and R. Jaeger, 149–165. New York: Kluwer Academic/Plenum.

Chippindale, P. T., A. H. Price, and D. M. Hillis. 1993. A new species of perennibranchiate salamander (*Eurycea*: Plethodontidae) from Austin, Texas. *Herpetologica* 49:248–259.

———. 1998. Systematic status of the San Marcos salamander, *Eurycea nana* (Caudata: Plethodontidae). *Copeia*, 1998: 1046–1049.

Chippindale, P. T., A. H. Price, J. J. Wiens, and D. M. Hillis. 2000. Phylogenetic relationships and systematic revision of Central Texas hemidactyline plethodontid salamanders. *Herpetological Monographs* 14:1–80.

Conant, R., and J. T. Collins. 1998. *A field guide to reptiles and amphibians of eastern and central North America.* Peterson Field Guide Series. Boston: Houghton Mifflin.

Crother, B. I., ed. 2008. *Scientific and standard English names of amphibians and reptiles of North America north of Mexico, with comments regarding confidence in our understanding.* 6th ed. Shoreview, Minn.: Society for the Study of Amphibians and Reptiles.

Dayton, G. H., R. Skiles, and L. Dayton. 2007. *Frogs and toads of Big Bend National Park.* College Station: Texas A&M Univ. Press.

Degenhardt, W. G., C. W. Painter, and A. H. Price. 1996. *Amphibians and reptiles of New Mexico.* Albuquerque: Univ. of New Mexico Press.

Dixon, J. R. 2000. *Amphibians and reptiles of Texas.* College Station: Texas A&M Univ. Press.

Duellman, W. E., and L. Trueb. 1994. *Biology of amphibians.* London: John Hopkins Univ. Press.

Dundee, H. A., D. A. Rossman, and E. C. Beckham. 1996. *The amphibians and reptiles of Louisiana.* Baton Rouge: Louisiana State Univ. Press.

Elliott, L., C. Gerhardt, and C. Davidson. 2009. *The frogs and toads of North America: A comprehensive guide to their identification, behavior, and calls.* Boston: Houghton Mifflin Harcourt.

Gascon, C., J. P. Collins, R. D. Moore, D. R. Church, J. E. McKay, and J. R. Mendelson III, eds. *Amphibian conservation action plan.* 2007. Gland, Switzerland: International Union for Conservation of Nature and Natural Resources. www.amphibianark.org/pdf /ACAP.pdf.

Hillis, D. M., D. A. Chamberlain, T. P. Wilcox, and P. T. Chippindale. 2001. A new species of subterranean blind salamander (Plethodontidae: Hemidactylini: Eurycea: Typhlomolge) from Austin, Texas, and a systematic revision of Central Texas paedomorphic salamanders. *Herpetologica* 57:266–280.

Hedges, S. B., Duellman, W. E., and Heinicke, M. P. 2008. New World direct-developing frogs (Anura: Terrarana): Molecular phylogeny, classification, biogeography, and conservation. *Zootaxa* 1737:1–182.

Lemos-Espinal, J. A., and H. M. Smith. 2007a. *Amphibians and reptiles of the state of Coahuila, Mexico.* Mexico City: UNAM-CONABIO.
———. 2007b. *Amphibians and reptiles of the state of Chihuahua, Mexico.* Mexico City: UNAM-CONABIO.

———. 2009. *Keys to the amphibians and reptiles of Sonora, Chihuahua, and Coahuila, Mexico.* Mexico City: UNAM-CONABIO.

López, L. O., G. A. Woolrich Pina, and J. A. Lemos Espinal. 2009. *La Familia Bufonidae en México.* Mexico City: UNAM-CONABIO.

Petranka, J. W. 1998. *Salamanders of the United States and Canada.* Washington, D.C.: Smithsonian Institution Press.

Pike, D. A., B. M. Croak, J. K. Webb, and R. Shine. 2010. Subtle—but easily reversible—anthropogenic disturbance seriously degrades habitat quality for rock-dwelling reptiles. *Animal Conservation* 13:411–418.

Reichling, S. B. 2008. *Reptiles and amphibians of the southern pine woods.* Gainesville: Univ. of Florida Press.

Sievert, G., and L. Sievert. 2005. *A field guide to Oklahoma's amphibians and reptiles.* Oklahoma City: Oklahoma Department of Wildlife Conservation.

Smith, H. M. 1978. *A guide to field identification amphibians of North America.* New York: Golden.

Stebbins, R. C., and N. W. Cohen. 1995. *A natural history of amphibians.* Princeton, N.J.: Princeton Univ. Press.

Trauth, S. E., H. W. Robison, and M. V. Plummer. 2004. *The amphibians and reptiles of Arkansas.* Fayetteville: Univ. of Arkansas Press.

INDEX OF COMMON NAMES

Bold page numbers indicate detailed discussion.

INDEX OF SCIENTIFIC NAMES